Are Your Diet Sodas Killing You?

Results from
My Aspartame Experiment

Victoria Inness-Brown, M.A.
aspartameexperiment.com

Writers Without Borders
DBA Writing by Design
Jamul CA 91935

Version 1.0

Library of Congress Cataloging-in-Publication Data
Victoria Inness-Brown, M.A.
Are Your Diet Sodas Killing You? Results from My Aspartame Experiment

ISBN 1456377736

1. Food. 2. Diet. 3. Nutrition. 4. Food additives. 5. Health. 6. Disease. 7. Pharmaceuticals. 8. Medical studies. 9. Politics.

December 2010

Cover and book design, photos by Victoria Inness-Brown, M.A.
Edited by Sally Altman, Damien Andrews, Gini Energy, Cher Gilmore, and Lalchumi Ralte

Printed in the United States of America
For additional copies, go to aspartameexperiment.com.

Contents

"Aspartame, because it is a poison that affects protein synthesis, because it affects how the synapse operates in the brain, and because it affects DNA, can affect numerous organs, so you can get a lot of different symptoms that seem unconnected."

"Again, it's this variability in your sensitivity to toxins. Some people may notice very little if anything. A majority of people will have one of a number of symptoms, because we know that the aspartame, because it is a poison that affects protein synthesis, because it affects how the synapse operates in the brain, and because it affects DNA, can affect numerous organs, so you can get a lot of different symptoms that seem unconnected."

—Dr. Russell Blaylock

This report documents an experiment I did as a private citizen, where I administered the artificial sweetener aspartame (in the form of NutraSweet mixed with water) to 60 rats (30 male, 30 female) between May 2002 and November 2004. Another 48 rats (24 male, 24 female) were used as controls.

> **Note: In November 2009, aspartame was rebranded AminoSweet by leading manufacturer, Ajinomoto. According to the company, "The name AminoSweet is appealing and memorable. It reflects that Amino-Sweet comes from the same amino acids that are abundant in the food we eat every day."**

Family members were heavy diet soda drinkers

Aspartame has been the subject of hundreds of experiments. So why did I undertake another one? Several years ago I became concerned about the health of family members who were drinking large amounts of diet soda containing aspartame. I was worried that they were consuming a chemical that could one day lead to their painful and early deaths—or worse, a walking death due to Alzheimer's, Parkinson's, or some other debilitating long-term illness brought on by aspartame.

Though this report does not focus on those neurological illnesses, neurosurgeon Russell Blaylock, MD, studied the causes of those diseases for over 20 years after his father passed on from Parkinson's. Blaylock's theories about neurotoxins such as aspartame and monosodium glutamate (MSG) are eloquently described and beautifully illustrated by Blaylock in his book, *Excitotoxins, the Taste that Kills.*

Convinced by the Bressler report

While researching problems associated with aspartame, I came across *The Bressler Report*, which was written by an auditor working for the U.S. Food and Drug Administration (FDA). The report describes the extensive analysis performed by Jerome Bressler, MD, on an aspartame-related study undertaken by G.D. Searle & Company, a pharmaceutical corporation founded in 1888 in Skokie, Illinois. Searle is the company that first marketed aspartame, and conducted the study using diketopiperazine (DKP), the most prevalent breakdown component of aspartame.

In his report, Dr. Bressler mentions numerous instances where Searle lacked forthrightness in reporting negative results to the FDA. For example, it states that tumors were removed and rats returned to the study without the tumors or surgery being reported to the FDA. One rat was even documented as dead, and then alive, and then dead again. Because most data related to Searle's aspartame studies was under FDA seal at Searle during the time of the Bressler audit, the fact that such incriminating information remained made me believe that Searle's real test results were far worse than the negative findings identified within *The Bressler Report*. While writing this report, I discovered that my belief appeared to be substantiated. According to the website, www.dorway.com:

"(Dr. Bressler) admitted the studies on aspartame were so bad that when his report was retyped, the FDA removed the worst 20 percent. So, as bad as this report is, it was originally worse."

"In a conversation with Dr. Betty Martini, Dr. H. J. Roberts and Dr. Russell Blaylock, (Dr. Bressler) admitted the studies on aspartame were so bad that when his report was retyped, the FDA removed the worst 20 percent. So, as bad as this report is, it was originally worse. Dr. Roberts wrote his congressman demanding the FDA release the other 20 percent, but they refused saying it was confidential."

I was struck by the number and size of the unreported growths identified in *The Bressler Report*, and became convinced that *I* might see tumors and possibly other adverse effects if I proceeded with my own aspartame experiment. After all, a 2-in. x 1.75 in. x 1 in. (5.0 cm x 4.5 cm x 2.5 cm) tumor would be hard to miss.

I wanted visual proof of aspartame's adverse effects

If I could provide visual proof of the adverse effects of aspartame, I thought it might convince my family and friends to avoid the chemical. As a technical writer, I find that visuals add power to the written word that can otherwise be easily manipulated with statistics, fabrications or high-tech mumbo-jumbo. I also felt that showing photos of adverse effects of aspartame might add credibility to the work of the undervalued individuals independently researching the adverse effects of aspartame, and add a new tool for those who have been fighting to get aspartame off the market.

For more information

For links to support groups, websites, videos, audios, and books related to aspartame, click the link For More Info at aspartameexperiment.com.

For more about the technical details of the protocol I used in my experiment and all references used in this report, refer to *My Aspartame Experiment: Report from a Private Citizen*. It is fully indexed and available with either full color or black and white photos from aspartameexperiment.com.

1

What is Aspartame?

"Aspartame is approximately 200 times sweeter than sugar, tastes like sugar, can enhance fruit flavors, saves calories and does not contribute to tooth decay. Products sweetened with aspartame can be useful as part of a healthful diet."

"Aspartame is approximately 200 times sweeter than sugar, tastes like sugar, can enhance fruit flavors, saves calories and does not contribute to tooth decay. Products sweetened with aspartame can be useful as part of a healthful diet."

—Aspartame Information Center, industry website

"Aspartame is a molecule composed of three ingredients, aspartic acid, 40% (an excitotoxin ... that stimulates the neurons of the brain to death), ... a methyl ester that immediately converts to methyl alcohol, (10%) then breaks down to formaldehyde (embalming fluid) and formic acid (ant sting poison), and 50% phenylalanine ... a neurotoxin that lowers the seizure threshold and depletes serotonin, triggering psychiatric and behavioral problems.... The FDA lists 92 documented symptoms from aspartame, from four types of seizures and coma, to male sexual dysfunction and death."

—Betty Martini, anti-aspartame activist

If you're unfamiliar with aspartame, now rebranded as AminoSweet, the following pages will introduce you to this ubiquitous man-made sweetener.

Chemical classification

Aspartame is classified as an artificial, synthetic, non-nutritive, non-caloric, low-cal, low-carb, diabetic-safe and alternative-sweetener-type food additive. It is ingested by an estimated 200 million people worldwide in more than 6,000 consumables, including diet sodas, fruit juices, candies, coffees, teas, pharmaceuticals, vitamins, and dairy products.

Brand names

Aspartame is marketed as a low-calorie tabletop sweetener under the names: NutraSweet,® NutraSweet Spoonfuls,® Equal,® Equal Measure,® Equal Sugar Lite, Canderel,® Mivida, Sweetex, Peptis, Natrasweet,® NatraTaste, Bienvia, Kroger Sweet Servings, Sugar Twin Plus, Twinsweet™, Miwon, Neotame, Sanecta, Tri-Sweet, E951, NouriSweet™, Aspartame, and most recently, AminoSweet.

Chemical description

The chemical structure of aspartame is shown in Figure 1-1.

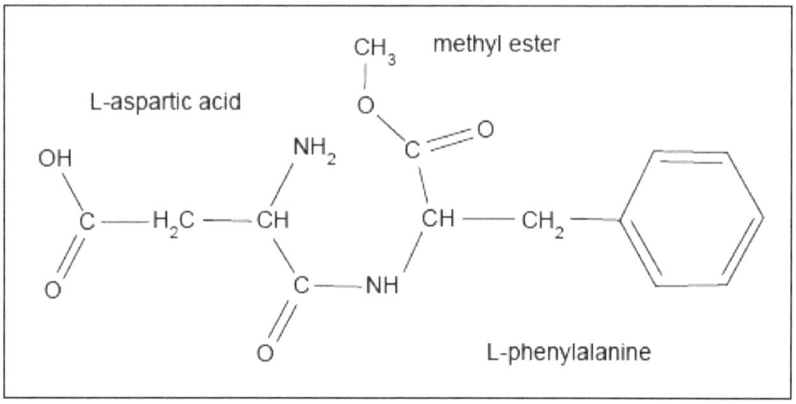

FIGURE 1-1: **Chemical structure of aspartame**

Scientifically speaking, aspartame is: N-L-alpha-aspartyl-L-phenylalanine 1-methyl ester.

Aspartame's scientific name is: *N-L-alpha-aspartyl-L-phenylalanine 1-methyl ester.* The name assigned to aspartame by the International Union of Pure and Applied Chemistry (IUPAC) is: *3-amino-4-(1-methoxycarbonyl-2-phenyl-ethyl) amino-4-oxo-butanoic acid.* Aspartame is considered chemically synonymous with many other chemical compounds, as listed in the endnotes of this chapter. Aspartame's Chemical Abstracts Service (CAS) identifier is: 22839-47-0. Its chemical formula is: $C_{14}H_{18}N_2O_5$, and its molecular mass is 294.30 g/mol.

As shown in Figure 1-1, the major chemical components of aspartame are:

- L-phenylalanine (CAS # 63-91-2), about 50%; a biosynthetic amino acid

- L-aspartic acid (CAS # 56-84-8), about 40%; a biosynthetic amino acid

- Methyl ester, about 10%; from treating L-phenylalanine with methanol (CAS # 67-56-1, wood alcohol), in the presence of hydrochloric acid.

2 My Aspartame Experiment

The Bressler Report motivated me to do my own rat study on the artificial sweetener aspartame from the end of May 27, 2002 through November 2004.

The Bressler Report motivated me to do my own rat study on the artificial sweetener aspartame, now rebranded as AminoSweet, from May 27, 2002 through November 2004. The report is FDA auditor Dr. Jerome Bressler's extensive analysis of an aspartame-related study done by G.D. Searle & Company, the original manufacturer of aspartame. For a detailed description of the chemical, see "What is Aspartame?" on page 1.

My experimental setup is shown in Figure 2-1. Not shown are the clay saucers and upside-down clay pots that served as the rats' houses.

FIGURE 2-1: **Experimental setup of my aspartame (AminoSweet) study**

My intent was to take video of any adverse effects, and to extract photos from the video.

My intent for the experiment was to take video of any adverse effects, and to extract photos from the video. The resulting photos appear throughout this report.

I attempted to purchase pure aspartame over the Internet, but the only website I found that sold it said that it was only available to food and beverage manufacturers. Since I was prevented from purchasing pure aspartame, I mixed packets of NutraSweet (Figure 2-2) in the water I gave the rats of my experimental group. The amount of aspartame my female rats received was equivalent to a 120 lb. (54.5 kg) human female drinking about 14 cans of diet soda per day. Or a 150 lb. (68 kg) male drinking about 13 cans per day. I have heard of people consuming more than this each day. One is Father Joe Carroll, President of the St. Vincent de Paul Village homeless assistance agency in San Diego. In November, 2008 Father Joe received an award at a meeting of the California Medical Board where I spoke briefly about my work. At that time, he said he drank about 20 diet sodas per day.

FIGURE 2-2: I mixed two packets of NutraSweet per each eight ounces of water I gave to the rats in my experimental group.

As my study progressed, I was amazed by the results. They contradict the studies done by the aspartame industry that consistently assure us of the safety of the additive, and instead confirm the results of independent investigators that consistently show the chemical to be hazardous.

For a searing analysis of the safety experiments performed by the aspartame industry, additional details about my experimental protocol, and all references used in this report, read the book *My Aspartame Experiment: Report from a Private Citizen*. It is available in full color or black and white from aspartameexperiment.com. For a shocking analysis of our food, water, and air that was instigated by online criticisms about my experiment, read Did I Fake My Data? in *My Aspartame Experiment: Report from a Private Citizen,* available from aspartameexperiment.com.

3

Resulting Tumors

"It is physiologically impossible for aspartame to cause cancer.... Long-term and lifetime tests in rats and mice with extremely large amounts of aspartame showed no evidence of brain tumors or any cancer associated with aspartame."

—International Food Information Council (IFIC) Foundation; *"IFIC is supported primarily by the broad-based food, beverage, and agricultural industries."*

"Aspartame ... was administered with feed to male and female Sprague-Dawley rats. Treatment lasted until spontaneous death of the animals.... The first results (show) that aspartame, in our experimental conditions, causes a statistically significant, dose-related increase in lymphomas...."

—Dr. Morando Soffritti, Fiorella Belpoggi, Davide Degli Esposti, and Luca Lambertini

Of all the effects of aspartame ingestion that I observed during my experiment, I found tumors to be the most frequent, often shocking, adverse effect. The sheer size and number of tumors in the aspartame group was astounding. I feel that there may be other reasons why the tumors were so plentiful and huge. For a discussion of my findings, see "Why so many tumors in my females on aspartame" in the book *My Aspartame Experiment: Report from a Private Citizen*, available from aspartameexperiment.com.

In his seven-year rat study on aspartame, Dr. Morando Soffritti from Bologna, Italy found aspartame to be associated with unusually high rates of lymphomas (cancer of the lymph glands), leukemias and other cancers. The study involved 1,900 laboratory rats that received doses of aspartame that started at rates equivalent to a 150-pound person drinking four to five 20-ounce bottles of diet soda per day.

In an article from the September 2007 issue of *Environmental Health Perspectives*, Soffritti states:

"In a previous study conducted at the Cesare Maltoni Cancer Research Center of the European Ramazzini Foundation (CMCRC/ERF), we demonstrated for the first time that aspartame (APM) is a multipotent carcinogenic agent when various doses are administered with feed to Sprague-Dawley rats from 8 weeks of age throughout the life span.

"[In a subsequent mega experiment] *we studied groups of 70-95 male and female Sprague-Dawley rats administered APM (2,000, 400, or 0 ppm) with feed from the 12th day of fetal life until natural death.*

"Our results show:

*"*a*) A significant dose-related increase of malignant tumor-bearing animals in males*

*"*b*) A significant increase in incidence of lymphomas/leukemias in males ... and a significant dose-related increase in incidence of lymphomas/leukemias in females*

*"*c*) A significant dose-related increase in incidence of mammary cancer in females*

"The results of this carcinogenicity bioassay confirm and reinforce the first experimental demonstration of APM's multipotential carcinogenicity at a dose level close to the acceptable daily intake for humans. Furthermore, the study demonstrates that when life-span exposure to APM begins during fetal life, its carcinogenic effects are increased."

In an apparent response to the damaging reports of the Soffritti studies published in 2005 and 2007, Ajinomoto of Japan—one of the world's largest producers of aspartame—sponsored its own safety evaluation of the artificial sweetener that has come to be known as the Magnuson 2007 review.

The conclusion of an article about the Magnuson 2007 review published in the journal *Critical Reviews in Toxicology* states that:

"Aspartame's metabolism is well understood and follows that of other common foods. Aspartame consumption, even at levels much higher than that expected under typical circumstances, has virtually no impact on levels of other blood constituents such as amino acids, methanol or glucose. Aspartame is a well-studied sweetener whose safety is clearly documented and well established through extensive laboratory testing, animal experiments, epidemiological studies, and human clinical trials. Controlled and thorough scientific studies confirm aspartame's safety and find no credible link between consumption of aspartame at levels found in the human diet and conditions related to the nervous system and behavior, nor any other symptom or illness. Aspartame is well documented to be nongenotoxic [non-toxic to genes] and there is no credible evidence that aspartame is carcinogenic. Aspartame does not increase hunger in those that use it; to the contrary, studies indicate it might be an effective tool as part of an overall weight management program. Aspartame is a well-characterized, thoroughly studied, high-intensity sweetener that has a long history of safe use in the food supply and can help reduce the caloric content of a wide variety of foods."

Health researcher Mark Gold, however, found massive conflicts of interest in those who performed the Magnuson 2007 review. According to Gold, in a well-documented investigative report online, *"Nearly every section of the Magnuson 2007 review has research that is misrepresented and/or crucial pieces of information are left out. In addition ... readers (including medical professionals) are often not told that this review was funded by the aspartame manufacturer, Ajinomoto, and the reviewers had enormous conflicts of interest."*

"The results of this carcinogenicity bioassay ... reinforce the first experimental demonstration of APM's multipotential carcinogenicity at a dose level close to the acceptable daily intake for humans."

Informa Healthcare, the parent company of the journal reporting the Magnuson 2007 study, stated in a press release that *"There were no known conflicts of interest with the sponsor* [Ajinomoto] *or potential biases of the authors."* Yet Gold found that most of the authors had serious conflicts of interest not documented in the report.

Gold notes that Informa Healthcare, the parent company of the journal reporting the Magnuson 2007 study, stated in a press release that *"There were no known conflicts of interest with the sponsor* [Ajinomoto] *or potential biases of the authors."* Yet Gold found most authors to have serious conflicts of interest not documented in the report. For example:

- Bernadine Magnuson: the lead author of the review (for whom the study was nicknamed the "Magnuson 2007 review") was the *"Senior Scientific and Regulatory Consultant for Cantox Health Sciences International, a corporate advocacy group. Cantox (now known as Intrinsik) specializes 'in assisting clients in their efforts to develop, gain regulatory approval and market products nationally or internationally.' ... In 2002, the president of Cantox, Ian C. Munro, worked directly with NutraSweet company employees and consultants on an aspartame review where he stated: 'After 30 plus years of rigorous scientific research, it is time to put questions of aspartame safety to rest.... The continuing debate over such a non-issue only serves to divert attention and the allocation of resources from more important health issues that need to be addressed.' Bernadene Magnuson became a member of the corporate advocacy group, The Burdock Group in 2005. The Burdock Group offers its clients 'technically rigorous, comprehensive safety and regulatory management of their products.... The Burdock Group offers the highest quality consulting services for the safety and regulatory issues facing the Food and Beverage, Dietary Supplement, Cosmetics/Personal Care and Pet Food Industries. Together, we form a cohesive team that offers single-source solutions for your business's safety assessment and regulatory needs.'"*

- G. A. Burdock: is part of the Burdock Group just mentioned.

- John Doull: *"a paid consultant of Monsanto* [previous manufacturer of aspartame], *a member of the Monsanto-funded ACSH Advisory Board, and a Trustee of the Monsanto- and Ajinomoto-funded corporate research association,* [International Life Sciences Institute] *ILSI."*

- R. M. Kroes: *"joined with Ian C. Munro, the president of the Cantox Health Sciences International corporate advocacy group, to work with Monsanto to review its herbicide, glyphosate."*

- G. M. Marsh: *"had research funded by the Formaldehyde Institute, a trade association consisting of Monsanto, Dupont and other chemical companies. The Formaldehyde Institute raised money for research in an attempt to portray formaldehyde exposure in a good light. Since independent published research has shown that aspartame ingestion leads to formaldehyde accumulation in the brain, kidneys, liver and other organs and tissues, Gary Marsh's research for the Formaldehyde Institute is a serious conflict of interest."*

- Michael Pariza: *"was a scientific advisor to the industry-funded advocacy group, 'American Council on Science & Health.' According to an article in the Washington Post: 'In 1982, the American Council on Science and Health (ACSH) filed a friend-of-the-court brief in a Formaldehyde Institute lawsuit that overturned a federal ban on formaldehyde insulation.... At least a third of ACSH's funding comes from such companies as Allied Corp., Coca-Cola, the National Soft Drink Association, Colgate-Palmolive Co., Dow Chemical Canada, Dupont, Eli Lilly, Exxon, General Mills, General Foods Fund, Gulf Oil, Hershey Foods, Johnson & Johnson, Kellogg's, Monsanto Fund, Mobil Foundation, M&M/Mars, Pillsbury Foundation, Procter & Gamble, Pfizer, Shell Oil, Upjohn and Velsicol Chemical.' Michael Pariza is also a member of the Board of Trustees of the International Life Sciences Institute (ILSI), a chemical and food company research association funded by Ajinomoto, Monsanto, Coca-Cola, PepsiCo, Nestle, and many other food and chemical companies involved in the production, use and sale of aspartame."*

- Ronald Walker: *"spent seven years as the ILSI's Chairman of their Scientific Committee on Toxicology/Food Safety in Europe. As mentioned earlier, ILSI is funded by Monsanto, Ajinomoto, Coca-Cola, Pepsi Cola, etc. He was a consultant for DSM Nutritional Products, a company that sold 'Twinsweet' from Holland Sweetener Company which is a mixture of aspartame and acesulfame-k. The DSM web site contained aspartame advocacy articles written by Holland Sweetener Company. He was a consultant for Numico Beheer BV/Danone Group, a company that had a joint venture with Ajinomoto (the sponsor of this review). He is a paid consultant to the corporate public relations group, the European Food Information Council with corporate members that include Coca-Cola, PepsiCo, Dannon, Nestle, etc. Finally, he was a paid consultant for Cantox Health Sciences International. Ronald Walker wrote a glowing review of another Ajinomoto product, monosodium glutamate (MSG) for a symposium funded by an Ajinomoto managed trade group, International Glutamate Technical Committee (IGTC). He participated in another aspartame review where he claimed that aspartame was safe."*

- Gary M. Williams: *"was the Chairman of the American Health Foundation (AHF) which was funded in part by The NutraSweet Company and other companies selling aspartame-containing products. AHF Board of Directors have included representatives of PepsiCo and the National Soft Drink Association. The AHF received more than $163,000 in grants from Philip Morris.... In 1987, the AHF convened a conference,* Sweeteners: Health Effects, *where an AHF representative concluded that aspartame and other sweeteners were safe: 'It is clear from the perspective of potential cancer risk that the sweeteners described in some detail in this report are safe and wholesome, and perhaps more so, than sugar. As we noted, it is our hope that this workshop will be the basis for international recognition of this fact, so that medical research effects can be directed effectively to areas more relevant to health maintenance.' Gary Williams also joined with Ian Munro to work with Monsanto to review its herbicide, glyphosate."*

Only two authors of the Magnuson 2007 review appear to be free of conflicts of interest, as Gold provided no evidence against them: P. S. Spencer from Oregon Health and Science University, Portland, Oregon, USA; and W. J. Waddell from the University of Louisville Medical School, Louisville, Kentucky, USA.

One of the studies mentioned in the Magnuson 2007 review was a massive study undertaken by the Division of Cancer Epidemiology and Genetics, Division of Cancer Control and Population Sciences, National Cancer Institute, National Institute of Health (NIH), Department of Health and Human Services (DHHS), Information Management Services, Inc., Rockville, Maryland; and the American Association of Retired Persons (AARP), Washington, District of Columbia.

According to this study:

"Background: In a few animal experiments, aspartame has been linked to hematopoietic and brain cancers. Most animal studies have found no increase in the risk of these or other cancers. Data on humans are sparse for either cancer. Concern lingers regarding this widely used artificial sweetener.

"Objective: We investigated prospectively whether aspartame consumption is associated with the risk of hematopoietic cancers or gliomas (malignant brain cancer).

"Methods: We examined 285,079 men and 188,905 women ages 50 to 71 years in the NIH-AARP Diet and Health Study cohort. Daily aspartame intake was derived from responses to a baseline self-administered food frequency questionnaire that queried consumption of four aspartame-containing beverages (soda, fruit drinks, sweetened iced tea, and aspartame added to hot coffee and tea) during the past year. Histologically confirmed incident cancers were identified from eight state cancer registries....

"Results: During over 5 years of follow-up (1995-2000), 1,888 hematopoietic cancers and 315 malignant gliomas were ascertained. Higher levels of aspartame intake were not associated with the risk of overall hematopoietic cancer (RR for 600 mg/d, 0.98; 95% CI, 0.76-1.27), glioma (RR for 400 mg/d, 0.73; 95% CI, 0.46-1.15; P for inverse linear trend = 0.05), or their subtypes in men and women.

"Conclusions: Our findings do not support the hypothesis that aspartame increases hematopoietic or brain cancer risk."

On the surface, the AARP and NIH study mentioned above looks impressive with 473,984 subjects. That is a massive number rarely seen in a single study. As you investigate it, however, you'll find it has major flaws. The biggest flaw is that it was based on a food survey that was not designed to evaluate aspartame consumption. Participants were only queried on the consumption *"of four aspartame-containing beverages (soda, fruit drinks, sweetened iced tea, and aspartame added to hot coffee and tea)."* Note that this description does not specify whether or not participants were questioned on their consumption of diet soda, diet fruit drinks, and artificially sweetened iced tea. It appears that those values were estimated. In addition, the study was highly limited regarding ingestion of aspartame, since the additive is present in over 6,000 consumables.

Regarding the validity of such a survey in general, the food frequency questionnaire itself has come into question. In another article from the same journal, Alan Kristal, Ulrike Peters, and John Potter made the statement:

"We are facing a crisis: hundreds of millions of dollars and many scientists' careers have been invested in studies using only FFQs to measure diet, but it is possible that these studies have not been, and will not be, able to answer many if not most questions about diet and cancer risk."

"Although painful to admit, it is possible that epidemiologists have been deluded in their acceptance of food frequency questionnaires (FFQ) as the standard tool for dietary assessment in large studies of diet and cancer. The substantial limitations of FFQs have been known for some time.... However, few of us expected the astonishingly poor measurement characteristics of FFQs ... nor had we expected to learn that diet and cancer associations detected when dietary assessment is based on dietary biomarkers or food records are undetectable when based on FFQs. We are facing a crisis: hundreds of millions of dollars and many scientists' careers have been invested in studies using only FFQs to measure diet, but it is possible that these studies have not been, and will not be, able to answer many if not most questions about diet and cancer risk."

In addition, the questionnaires of many people who may have had blood or brain cancers were thrown out of the study. From an article about the NIH-AARP Diet and Health Study we find: *"Out of 617,119 questionnaires returned, 567,169 were satisfactorily completed. We excluded ... 582 persons who had died or moved out of the study area before study entry, 52,887 persons with history of cancer ... or with only death reports of cancer."* So the questionnaires of many of the people who may have had cancers due to aspartame ingestion were eliminated from the study.

Tumors in my aspartame group

My females on aspartame were hardest hit, with a phenomenal 20 out of 30 = two-thirds or 67% of them developing visible tumors as shown in the following subsection. For an in-depth analysis of possible causes, see "Why so many tumors in my females on aspartame" in *My Aspartame Experiment: Report from a Private Citizen*, available from aspartameexperiment.com. In comparison, only 7 out of 30 = 23% of my males on aspartame developed visible tumors. See "Males on aspartame with tumors" on page 20. For observations about these results, see "Females on aspartame had three times more tumors than males" in *My Aspartame Experiment: Report from a Private Citizen*, available from aspartameexperiment.com.

Females in my control group also developed tumors. See "Tumors in my control group" on page 24. For a frightening analysis of possible causes, see "What may have caused my control group tumors?" in *My Aspartame Experiment: Report from a Private Citizen*, available from aspartameexperiment.com.

For a complete summary of my results, see "Summary of my experimental results" on page 47.

Females on aspartame with tumors

The following subsections show various groups of females on aspartame with tumors, organized by color.

White and black females with lymph or mammary gland tumors

The females in my experiment shown in Figure 3-1 through Figure 3-3 developed apparent tumors of the lymph or mammary gland. According to a local veterinarian, when a rat develops a tumor, it is most likely malignant. According to the book, *The Rat: An Owner's Guide to a Healthy Pet*, however, "*Most growths that rats get are benign.*" Since I did not have necropsies performed, we don't know which tumors were benign or malignant. However, I would not want to walk around with any of these tumors, whether benign or not.

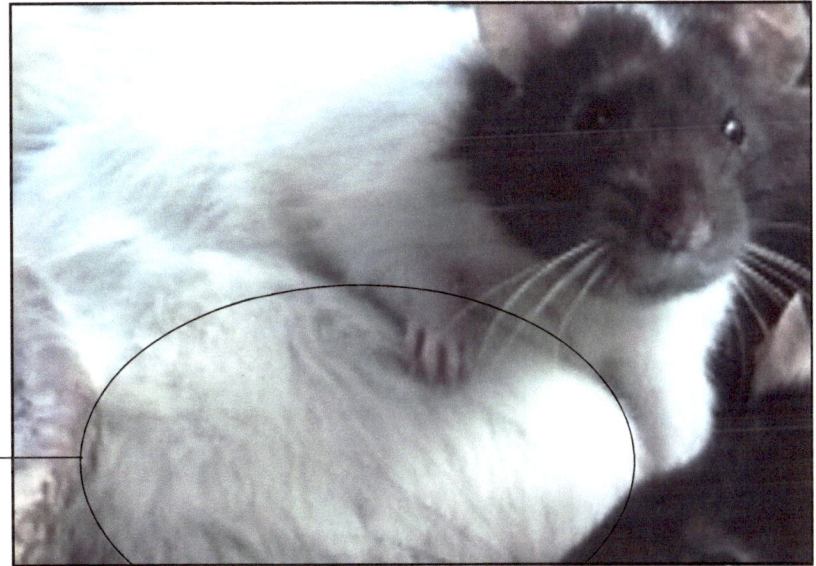

This female on aspartame often used her tumor as a pillow.

FIGURE 3-1: Female on aspartame that developed a huge tumor of the breast or lymph gland that she often used as a pillow.

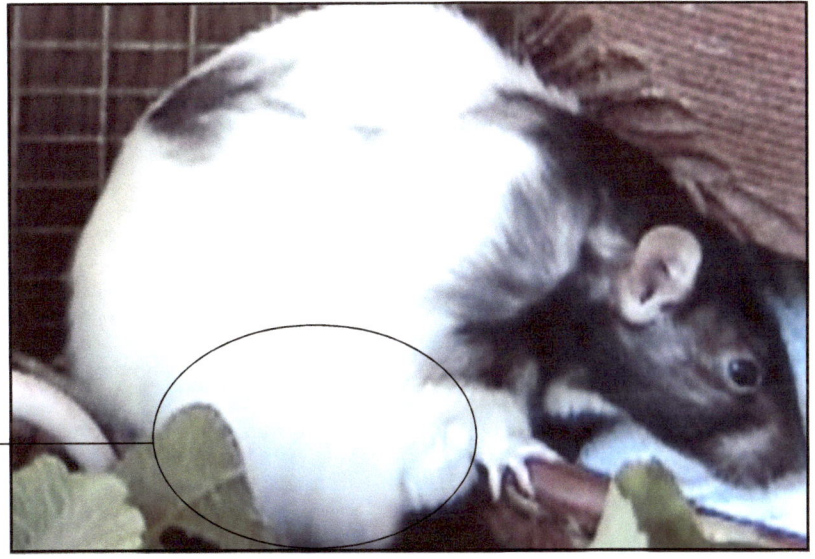

This female had an apparent lymph or mammary gland tumor on her right side.

FIGURE 3-2: Female on aspartame with an apparent lymph or mammary gland tumor.

The female in Figure 3-3 also developed an apparent lymph or mammary gland tumor. She was one of the last surviving rats of the experiment ending in November 2004.

This female had an apparent lymph or mammary gland tumor near her right foreleg.

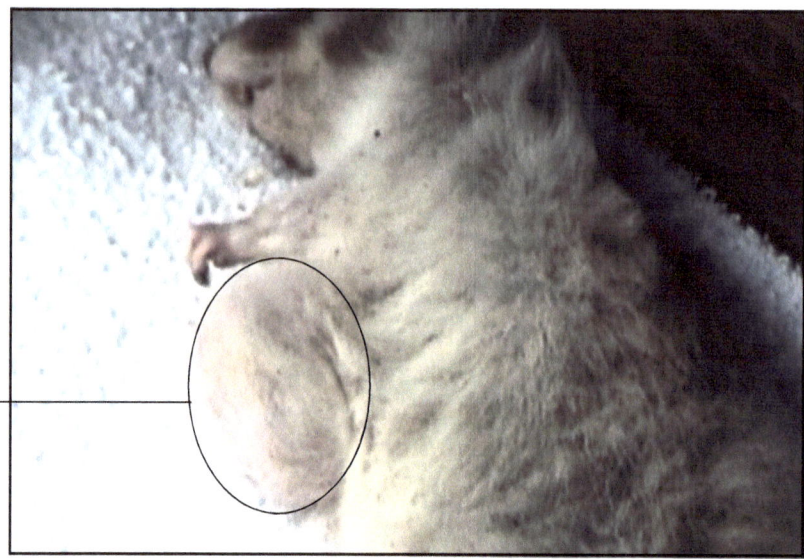

FIGURE 3-3: **Female on aspartame with an apparent lymph or mammary gland tumor.**

White and black females with lymph or mammary gland tumors and yellow tinged fur

In my experiment, the white and black females on aspartame shown in Figure 3-4 through Figure 3-14 had apparent lymph or mammary gland tumors. In addition, each had tinges of yellowing fur, which unfortunately is not discernible in the black-and-white version of this report. It is, however, a possible effect of aspartame. (See "Yellowing fur" on page 44.)

This female had an apparent lymph or mammary gland tumor near her left foreleg.

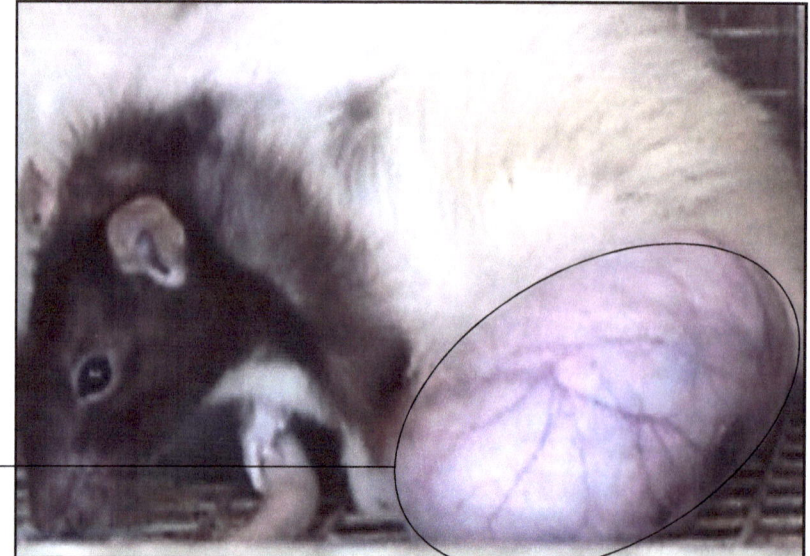

FIGURE 3-4: **Female on aspartame with an apparent lymph or mammary gland tumor and yellowing fur.**

The female in Figure 3-4 also had skin problems, indicated by the scabbing which is barely visible on the far-right underside of her tumor. (See "Skin disorders within my aspartame group" on page 39.) The tumor on the female in Figure 3-5 grew to gigantic proportions and appeared filled with numerous smaller tumors. Note how it almost completely surrounded her left foreleg, causing her great difficulty when walking.

The tumor on this female grew to enormous proportions and appeared filled with numerous smaller tumors.

Note how the tumor sack almost completely engulfed her left foreleg.

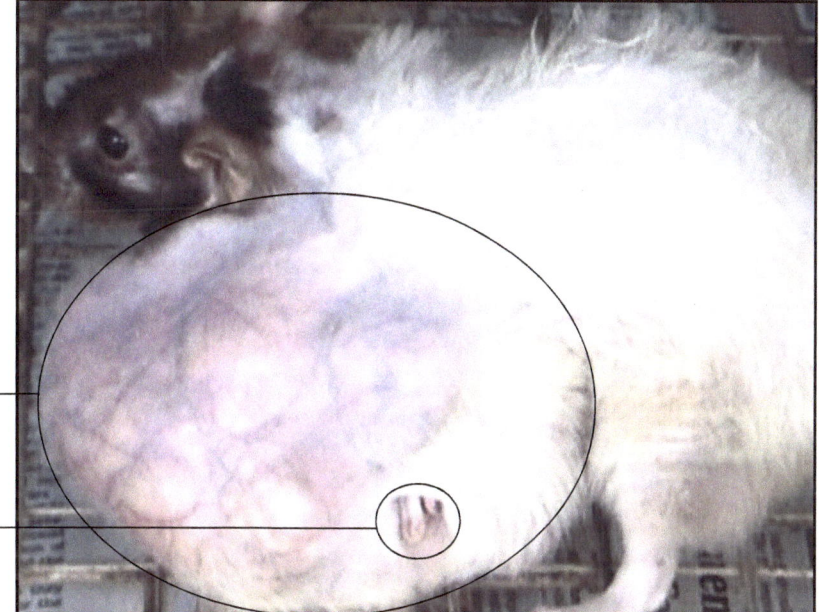

FIGURE 3-5: Female on aspartame that developed an immense lymph or mammary growth that appears subdivided into smaller tumors.

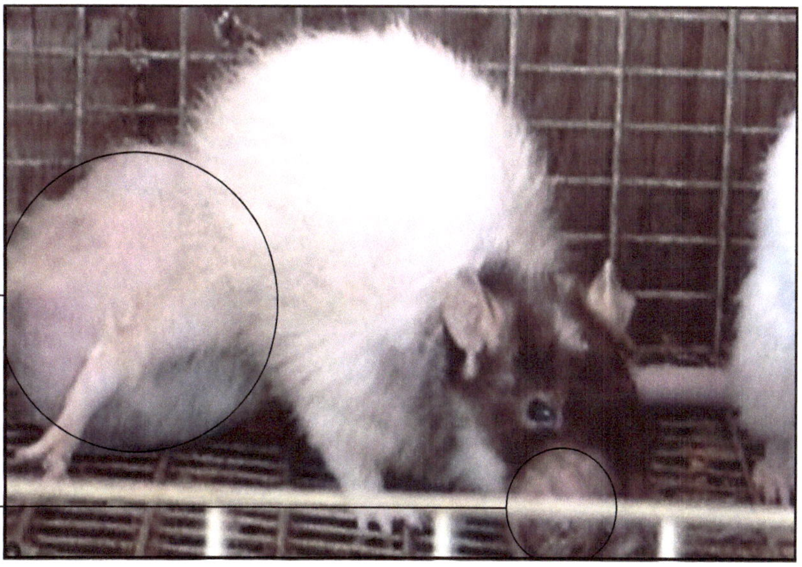

This female had an apparent lymph or mammary gland tumor near her right hind leg.

She also had a tumor on the right side of her face.

FIGURE 3-6: Female on aspartame with a tumor of the lymph or mammary gland and another on her face.

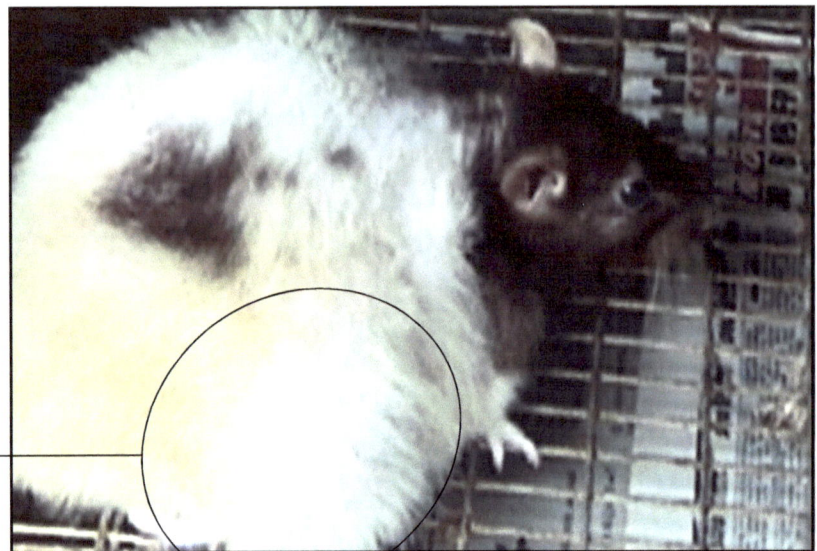

This female had an apparent lymph or mammary gland tumor near her right foreleg.

FIGURE 3-7: Aspartame female with an apparent lymph or mammary gland tumor near her right foreleg.

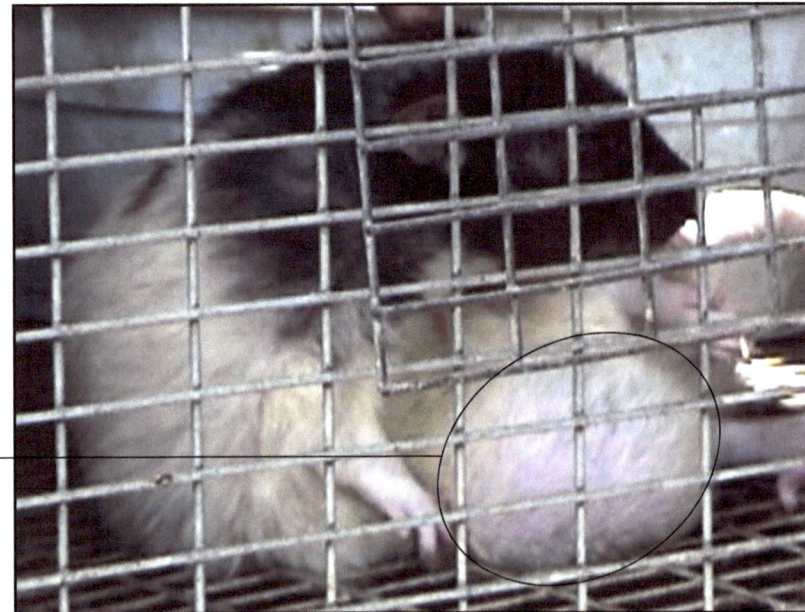

This female had an apparent tumor of the lymph or mammary gland near her right hind leg.

She also appeared to have bleeding eyes.

FIGURE 3-8: Aspartame female with an apparent tumor of the lymph or mammary gland and bleeding eyes.

For a discussion of aspartame and bleeding eyes, see "Bleeding eyes" on page 35.

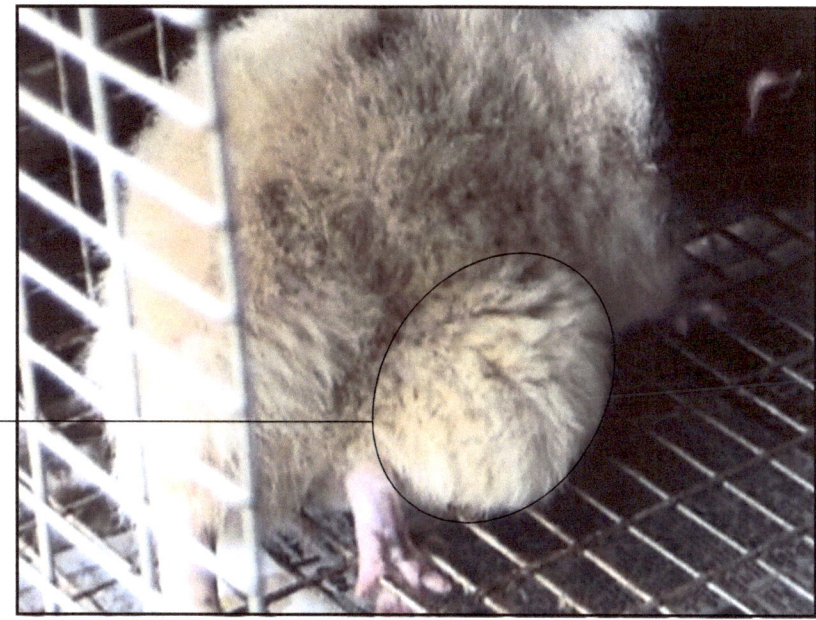

This female appeared to have a lymph or mammary gland tumor in front of her right hind leg.

FIGURE 3-9: **Female on aspartame with an apparent lymph or mammary gland tumor.**

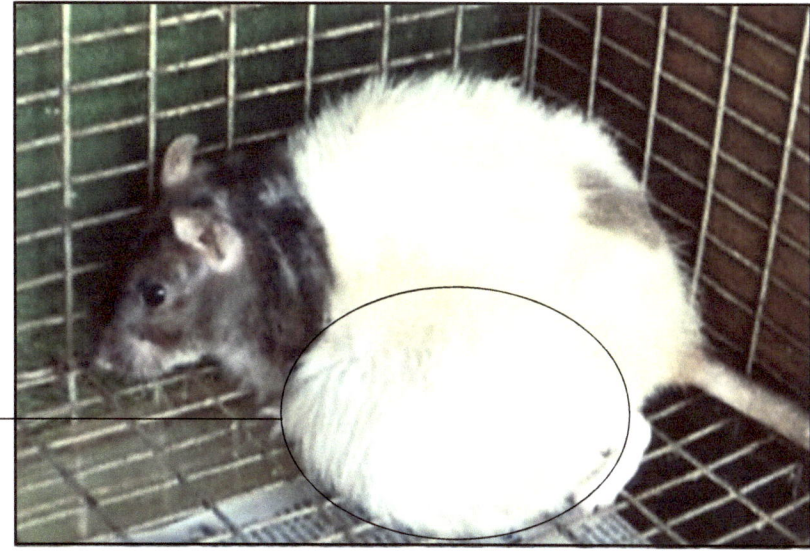

This female had an apparent tumor of the lymph or mammary gland on her left side.

FIGURE 3-10: **Female on aspartame with an apparent lymph or mammary gland tumor.**

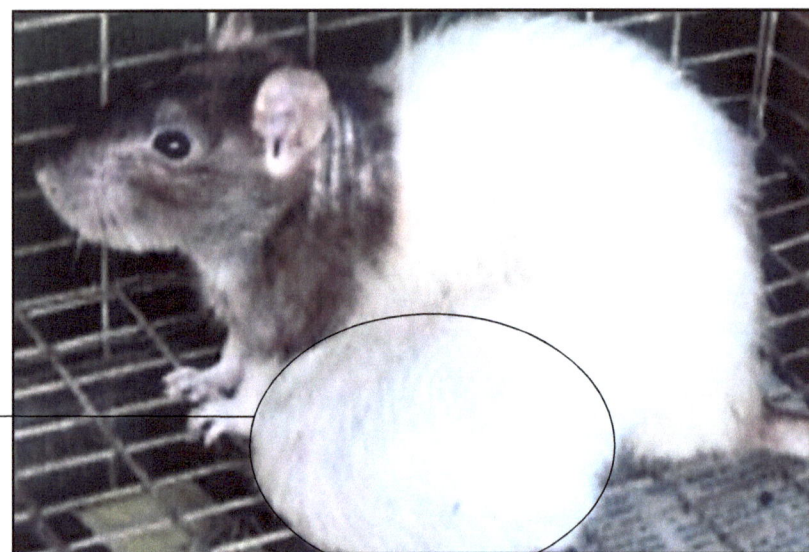

This female had an apparent tumor of the lymph or mammary gland on her left side.

FIGURE 3-11: **Female on aspartame with an apparent lymph or mammary gland tumor.**

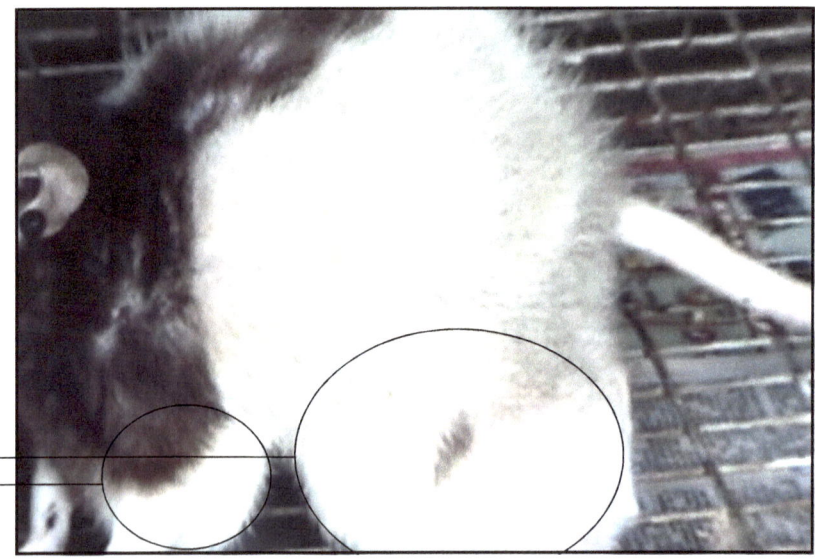

This female had multiple tumors of the lymph or mammary gland on her left side.

FIGURE 3-12: **Female on aspartame with an apparent lymph or mammary gland tumor.**

This female developed an apparent lymph or mammary gland tumor that almost completely engulfed her left foreleg.

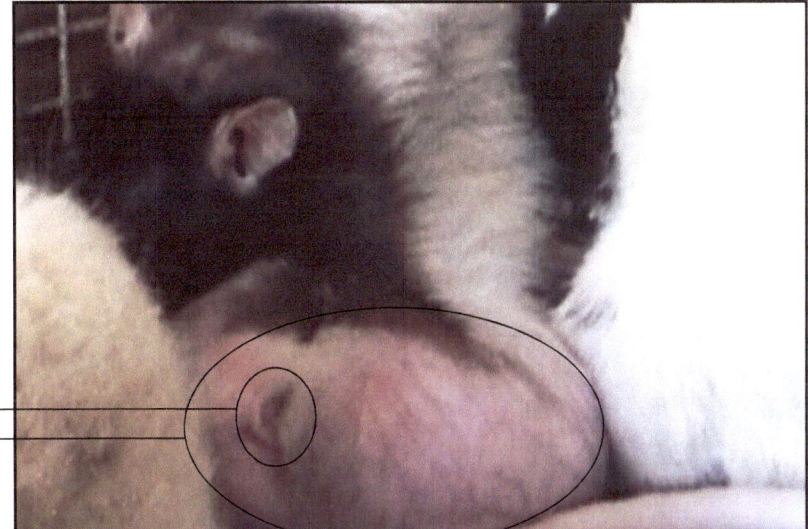

FIGURE 3-13: Female on aspartame with an apparent lymph or mammary gland tumor.

The female in Figure 3-14 appears to have a tumor of the lymph gland on the left side of her neck.

She also had protruding eyes, a symptom in humans of the thyroid disorder called Grave's disease (see "Protruding eyes" on page 37).

This female appeared to have a lymph gland tumor on the left side of her neck.

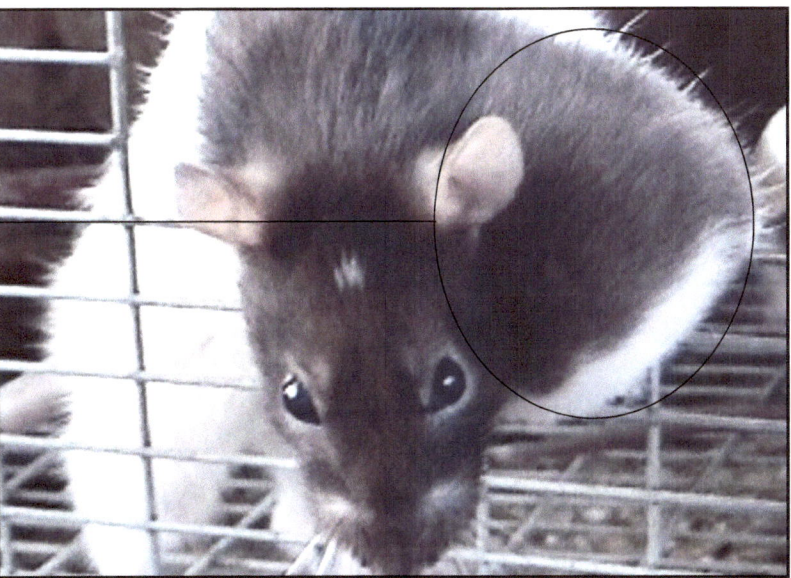

FIGURE 3-14: Female on aspartame with an apparent lymph gland tumor and protruding eyes.

Black and brown females with lymph or mammary gland tumors

In my experiment, the brown and black females on aspartame shown in Figure 3-15 through Figure 3-19 had apparent lymph or mammary gland tumors.

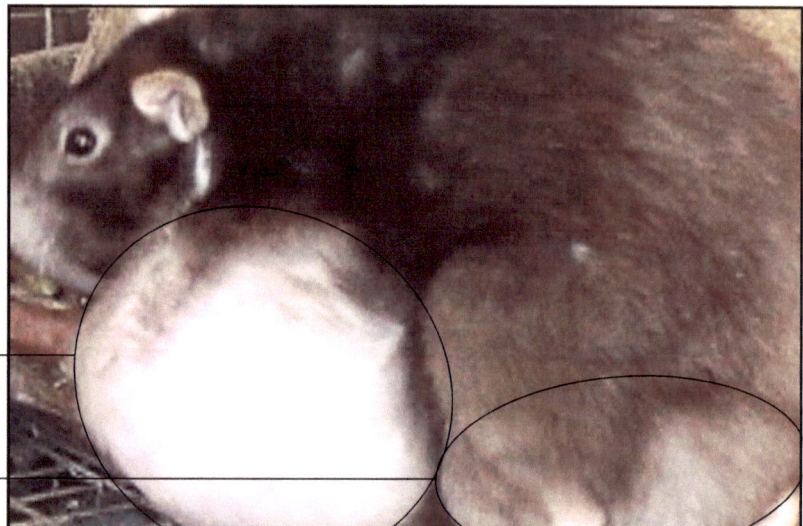

This female developed an apparent lymph gland tumor and a secondary lymph or mammary gland tumor.

FIGURE 3-15: Female on aspartame with an apparent lymph or mammary gland tumor.

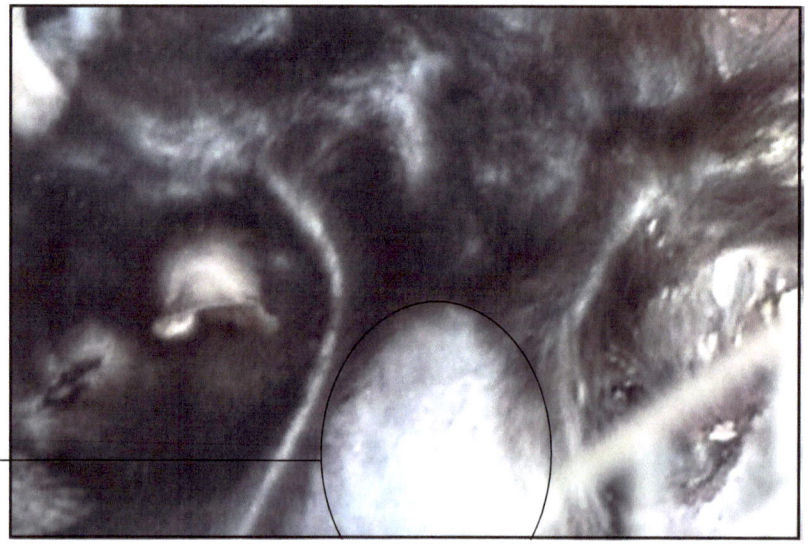

This female appeared to have a lymph or mammary gland tumor behind her left foreleg. She also had thinning fur and skin problems.

FIGURE 3-16: Female on aspartame with an apparent lymph or mammary gland tumor.

The female in Figure 3-16 also had thinning fur and developed severe skin problems as shown in Figure 4-18 (see "Thinning fur" on page 42 and "Skin disorders within my aspartame group" on page 39).

This female appeared to have a lymph or mammary gland tumor on her right side.

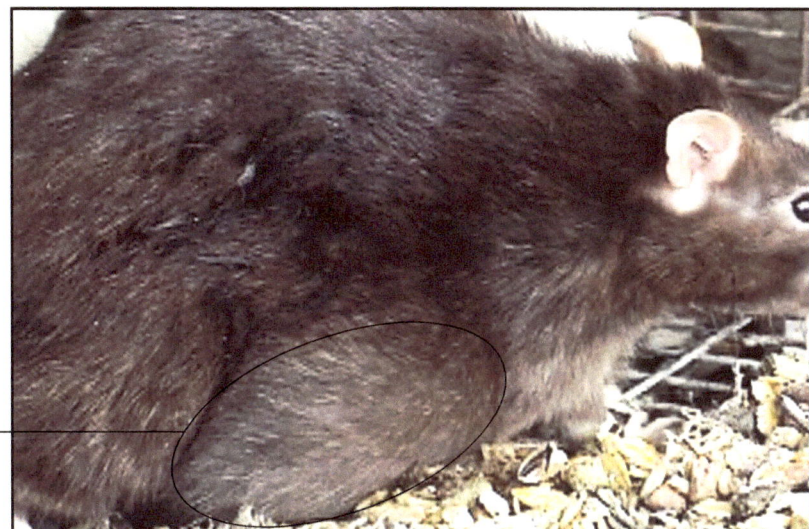

FIGURE 3-17: Female on aspartame with an apparent lymph or mammary gland tumor.

This female appeared to have a lymph or mammary gland tumor on her left side.

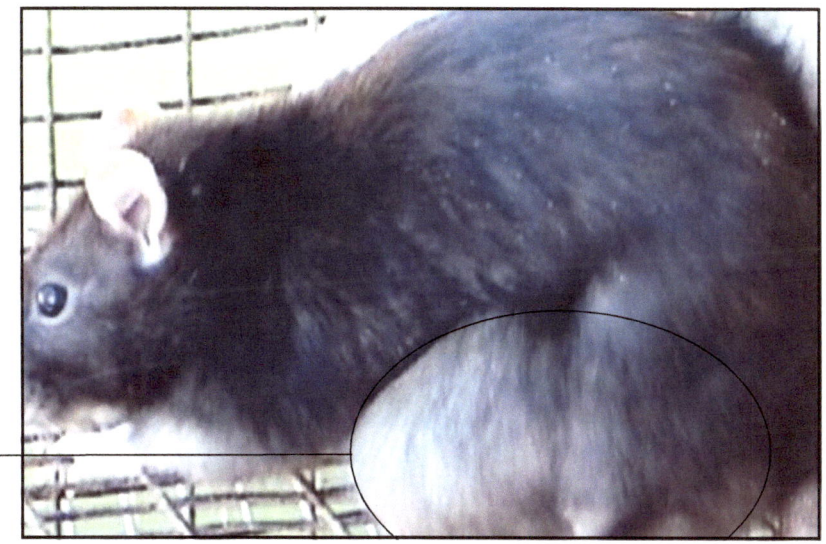

FIGURE 3-18: Female on aspartame with an apparent lymph or mammary gland tumor.

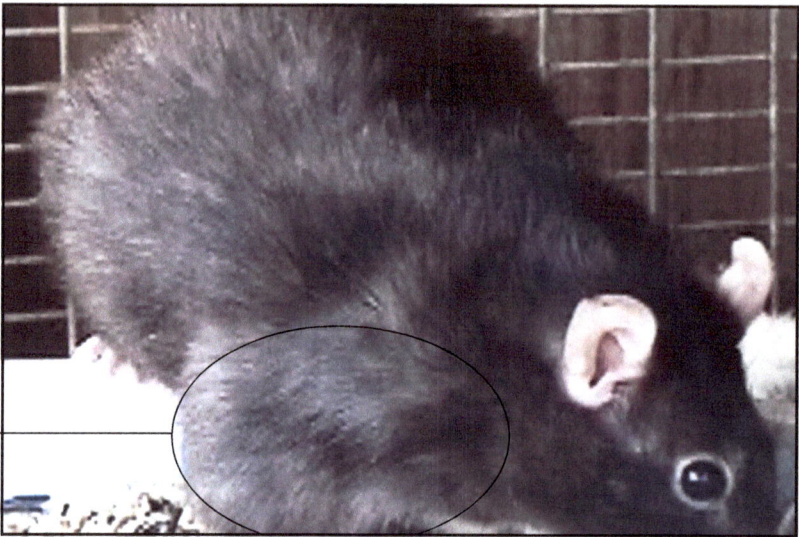

This female appeared to have a lymph or mammary gland tumor behind her right foreleg.

FIGURE 3-19: Female on aspartame with an apparent lymph or mammary gland tumor.

Beige female with lymph or mammary gland tumors

In my experiment, the beige female on aspartame shown in Figure 3-20 developed multiple lymph or mammary gland tumors.

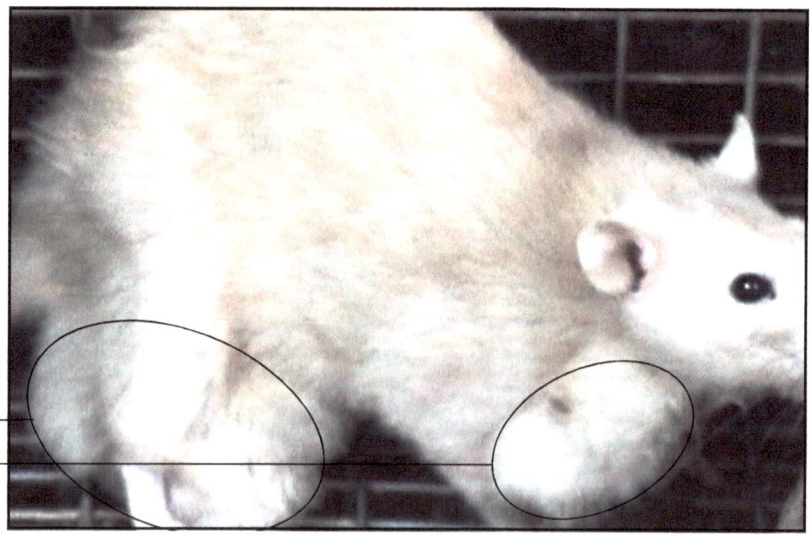

This female appeared to have multiple lymph or mammary gland tumors.

FIGURE 3-20: Female on aspartame with multiple lymph or mammary gland tumors.

Males on aspartame with tumors

The males on aspartame shown in Figure 3-21 through Figure 3-27 had apparent mammary or lymph gland tumors.

Those in Figure 3-21 and Figure 3-23 also appeared to have tinges of yellowing fur, a possible symptom of aspartame poisoning (see "Yellowing fur" on page 44).

This male had a growth on his belly that was possibly a mammary or lymph gland tumor. It made it difficult for him to maintain an upright position.

He also had yellowing fur, a possible symptom of aspartame poisoning.

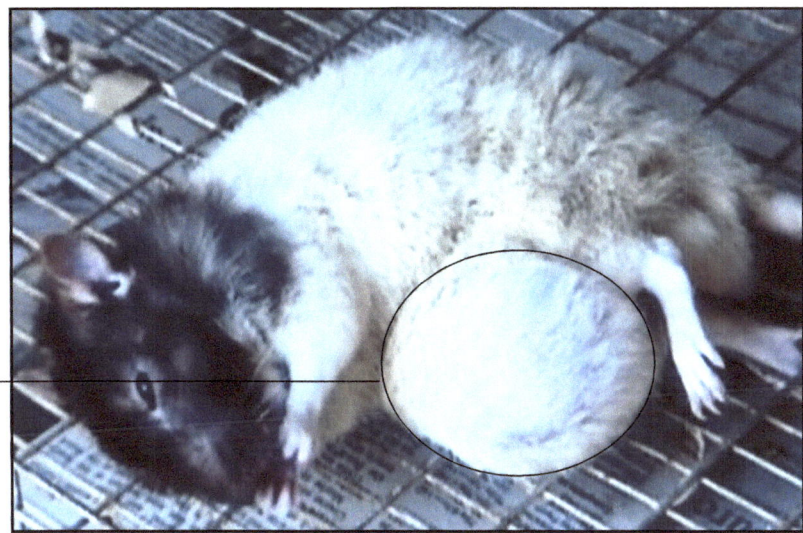

FIGURE 3-21: **Male on aspartame with an apparent mammary or lymph gland tumor near his left hind leg.**

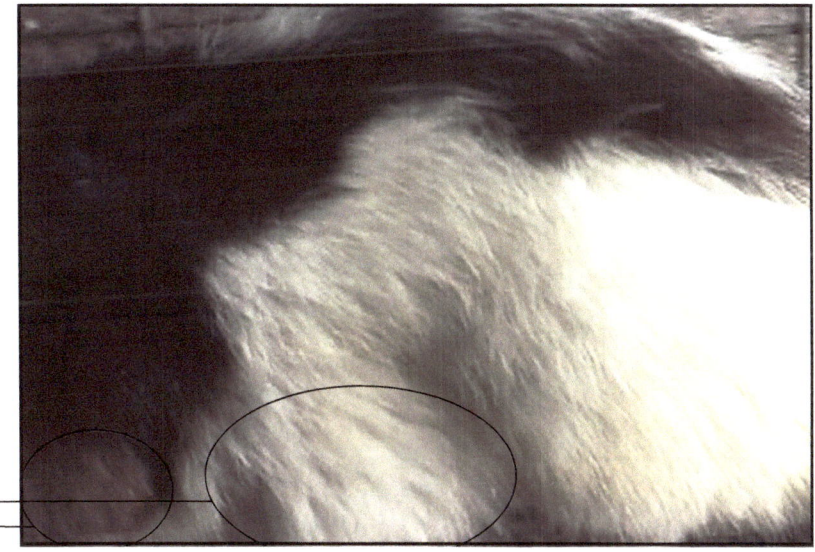

This male had multiple growths near his left foreleg which were apparent mammary or lymph gland tumors.

FIGURE 3-22: **Male on aspartame with an apparent lymph gland tumor in front of and mammary or lymph gland tumor behind his left foreleg.**

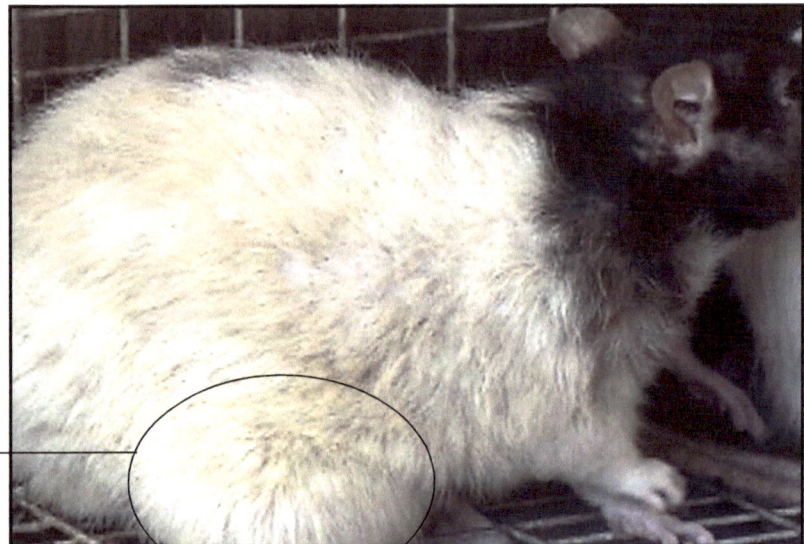

This male had a growth near his right hind leg that appears to have been a mammary or lymph gland tumor.

FIGURE 3-23: Male on aspartame with an apparent mammary or lymph gland tumor above his right hind leg.

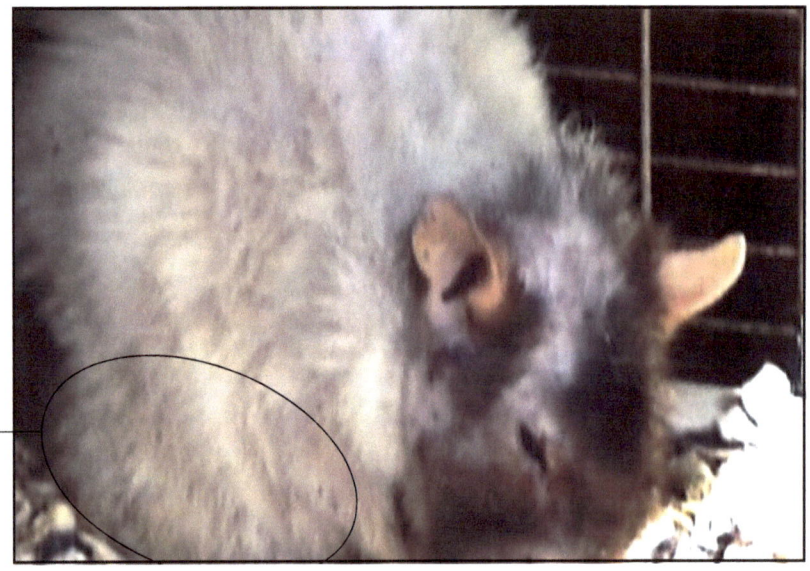

This male had a growth near his right foreleg that appears to have been a mammary or lymph gland tumor.

FIGURE 3-24: Male on aspartame with an apparent mammary or lymph gland tumor near his right foreleg.

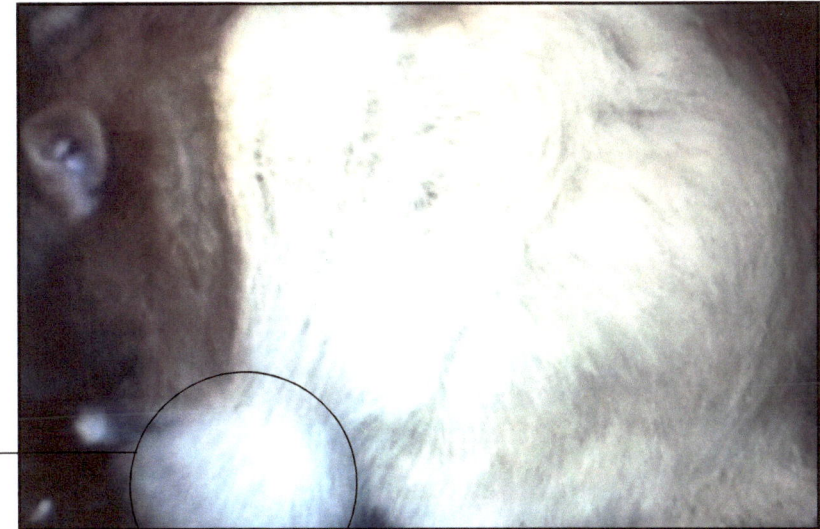

This male had a growth near his left foreleg which appears to have been a mammary or lymph gland tumor.

FIGURE 3-25: Male on aspartame with an apparent mammary or lymph gland tumor near his left foreleg.

The male in Figure 3-25 also grew obese (see "Obesity within my aspartame group" on page 45).

The male in Figure 3-26 had a growth on the left side of his face, which diminished in size after he got into a fight. For more about aspartame and aggressive behavior, see the discussion on page 34.

This male had a growth on the left side of his face.

FIGURE 3-26: Male on aspartame who developed a tumor on the left side of his face.

The male on aspartame shown post mortem in Figure 3-27, had a tumor on the right side of his face. I found him a few days after he died hidden behind one of the clay houses in his cage.

This male had a growth on the right side of his face. He was one of the first casualties of my experiment. He died within weeks after it started.

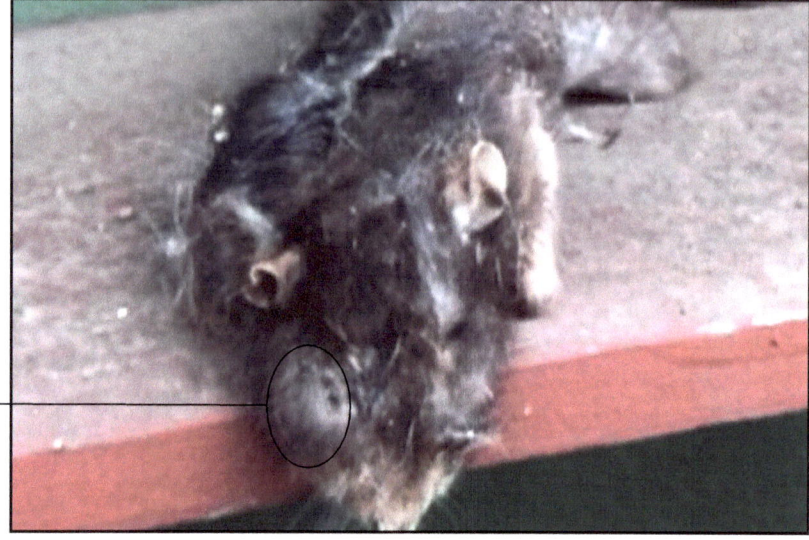

FIGURE 3-27: **Male on aspartame who developed a tumor on the right side of his face.**

Tumors in my control group

In my experiment, the control females shown in Figure 3-28 through Figure 3-31 developed apparent lymph or mammary gland tumors.

This control female developed an apparent lymph or mammary gland tumor on her left side.

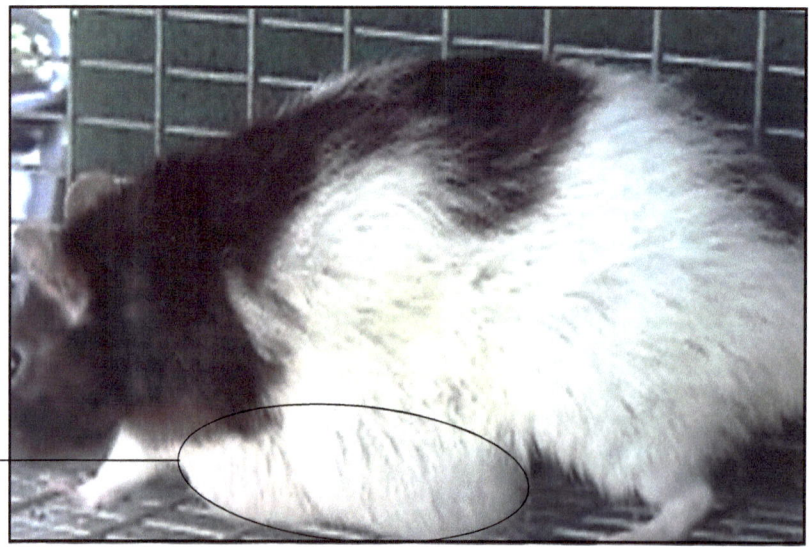

FIGURE 3-28: **Female from the control group with an apparent lymph or mammary gland tumor.**

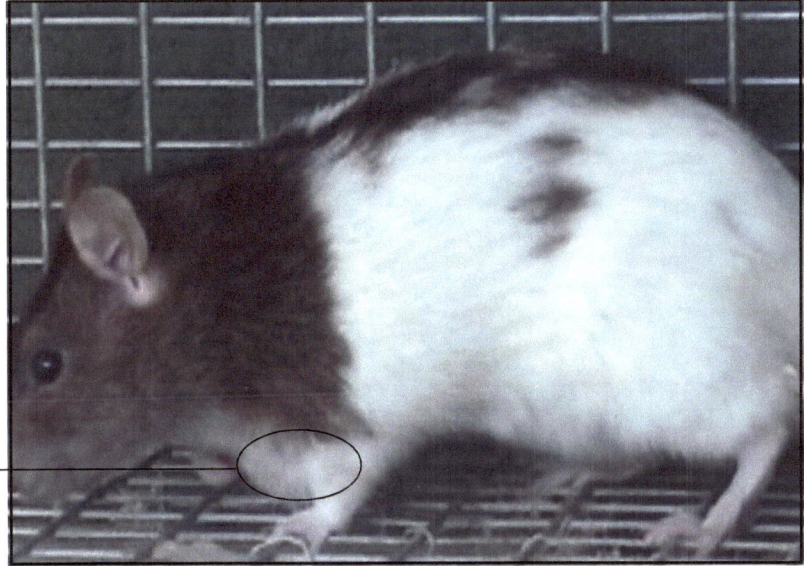

This control female developed an apparent lymph or mammary gland tumor near her left foreleg.

FIGURE 3-29: **Control group female with apparent lymph or mammary gland tumor.**

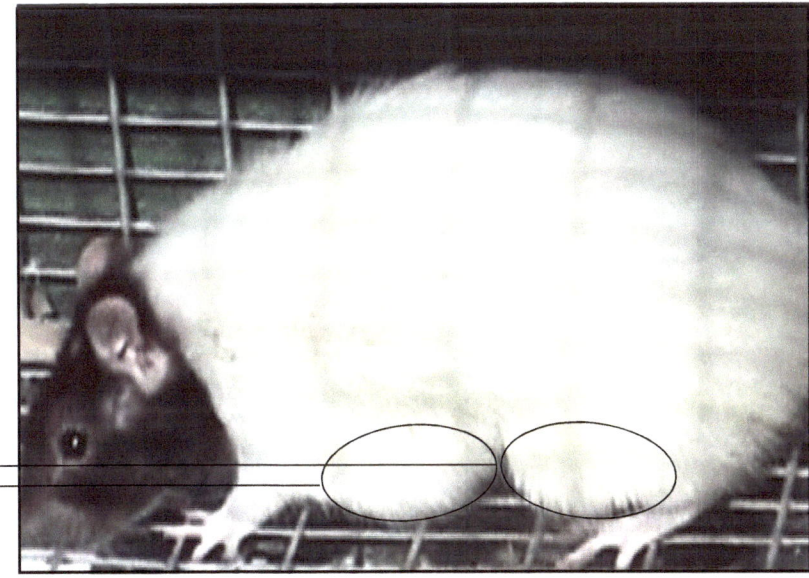

This control female developed multiple lymph or mammary gland tumors on her left side.

FIGURE 3-30: **Control group female with multiple lymph or mammary gland tumors.**

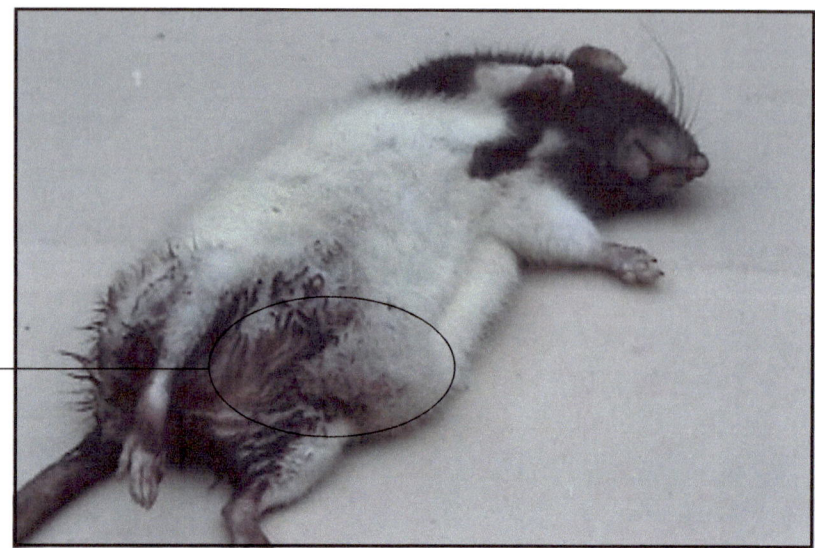

This control female developed an apparent lymph or mammary gland tumor on her left side.

FIGURE 3-31: White and black control group female with an apparent lymph or mammary gland tumor.

The brown female shown in Figure 3-32 developed an apparent lymph or mammary gland tumor under her right foreleg.

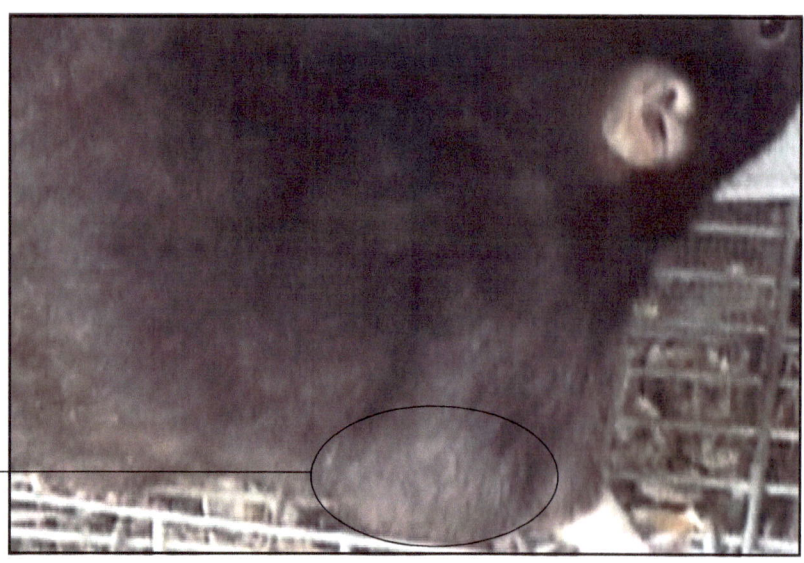

This control female developed an apparent lymph or mammary gland tumor behind her right foreleg.

FIGURE 3-32: Brown control group female with an apparent lymph or mammary gland tumor.

For a summary of the results shown in this chapter, see "Conclusion" on page 47.

4

Other Adverse Results

"Let's just have a quick review of what MSG and the excitotoxins [such as aspartame] *do. Well, they are associated with neurodegenerative diseases, they are associated with neurodevelopmental abnormalities, nervous system injury ... endocrine disorders, diabetes Types 1 and 2, Syndrome X, gross obesity, enhanced cancer growth and spread, immune dysfunction, retinal disorder, arterial sclerosis, multiple sclerosis, lupus and other auto-immune disorders, GI disorders and sudden cardiac death."*

—Dr. Russell Blaylock, neurosurgeon

The above quote exposes the wide range of adverse effects that may be associated with aspartame, now rebranded as AminoSweet. The tumors I observed are shown in Chapter 3. The following subsections show the diversity of additional adverse effects I observed:

- ◼ "Neurological disorders within my aspartame group" on page 27

- ◼ "Eye disorders within my aspartame group" on page 32

- ◼ "Skin disorders within my aspartame group" on page 39

- ◼ "Thinning & yellowing fur within my aspartame group" on page 42

- ◼ "Obesity within my aspartame group" on page 45

Neurological disorders within my aspartame group

Does aspartame cause neurological disorders? In the Index of *Aspartame Disease, an Ignored Epidemic*, Dr. Roberts lists the following "neurologic complications" that he observed in his patients as possibly being associated with aspartame consumption:

"Alzheimer's, amyotrophic lateral sclerosis (ALS), attention-deficit disorder and hyperactivity, carpal tunnel syndrome, cataplexy, confusion, dizziness, dopa-responsive dystonia, facial pain, hypnagogic hallucinations, intellectual deterioration, memory loss, motor neuron disease, muscle weakness, myasthenia gravis, neuralgia, Parkinson's disease, peripheral neuropathy, pseudotumor cerebri, restless legs syndrome, Sjogren's syndrome, sleep apnea, sleep paralysis, slurring of speech, multiple sclerosis, Tourette's syndrome, tremors, unexplained blackouts, unexplained pain, unsteadiness."

The following subsections show images of my rats on aspartame with visible symptoms of various types of neurological disorders.

Paralysis

"I have been seeing doctors for over three years now trying to find out why I am being turned into a cripple. I have severe weakness in upper legs and upper arms, they have tested me for everything under the sun and keep telling me I do not have anything they recognize.... I started to notice that my bad cycles coincided with how much diet soft drink I was using at the time."

During my experiment, the hind legs of the male on aspartame shown in Figure 4-1 became paralyzed.

The hind legs of this male on aspartame became paralyzed.

FIGURE 4-1: The hind legs of this male on aspartame became paralyzed

ALS, also known as Lou Gehrig's disease, can lead to paralysis. According to Dr. Blaylock, *"There is growing evidence that excitotoxins* [such as aspartame] *play a major role in a whole group of degenerative brain diseases in adults—especially the elderly. These diseases include Parkinson's disease, Alzheimer's disease, Huntington's disease.... ALS, as well as several more rare disorders of the nervous system."*

Spasmodic torticollis

Figure 4-2 shows a female on aspartame who continually twisted her head to her left, similar to the human disease, idiopathic spasmodic torticollis (IST). eMedicine states that: *"Torticollis is a condition that causes the neck to involuntarily twist to one side secondary to contraction of the neck muscles. The ear is tilted toward the contracted muscle and the chin is facing the opposite direction."* It is interesting to note that eMedicine states that torticollis may be chemically induced.

The neck of this female on aspartame was continually twisted to the left.

In humans, this is a symptom of the neurological disorder spasmodic torticollis, also referred to as dystonia.

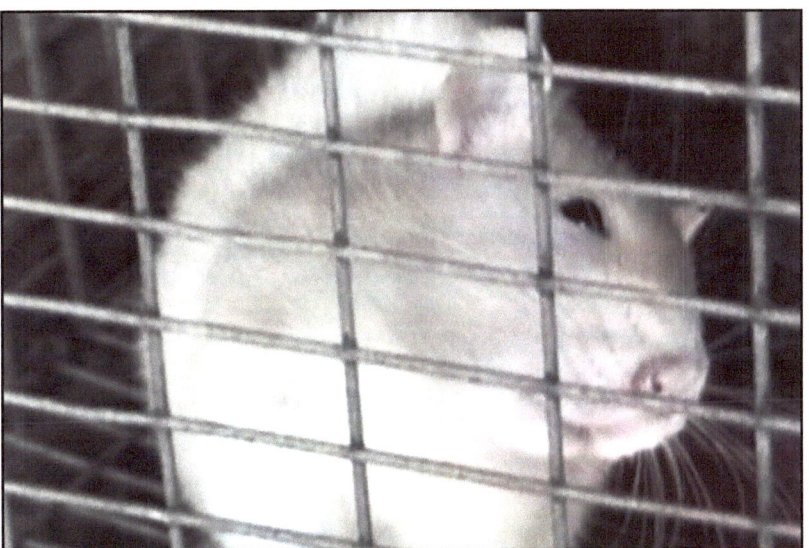

FIGURE 4-2: This female on aspartame appeared to have torticollis

In *Aspartame Disease, an Ignored Epidemic*, H.J. Roberts, MD, discusses dopa-responsive dystonia as being affected by the ingestion of aspartame because of its breakdown component phenylalanine: "*Hereditary progressive dystonia is another neurologic disorder that usually affects children.... Patients with this condition (at times misdiagnosed as cerebral palsy) can be adversely affected by ingesting phenylalanine because of decreased hepatic phenylalanine hydroxylase activity.*"

Cerebral palsy

"*During my pregnancy with my nine year old, I consumed sugar-free frozen yogurt (I craved it!) and he was born four months early, weighed one pound, nine ounces and now suffers from cerebral palsy and mental retardation. Until now, we all assumed that these were both due solely to premature birth. Now I wonder if he was born prematurely because I consumed aspartame and if so, did the aspartame cause all of this?*"

I was unable to capture it on camera, but one of my females continually moved her head from side to side, a possible symptom of cerebral palsy.

According to aspartame investigator and victim Jim Bowen, MD, and former FDA investigator Arthur M. Evangelista, "*During maternal aspartame consumption, development of the fetal nervous system is damaged or impaired via excitotoxic-saturated placental blood flow that can cause or contribute to cerebral palsy and pervasive developmental disorders....*

"*This is due to an incompetent blood brain barrier and neuronal (brain) damage produced by excitotoxins circulating in the fetal brain areas. This is especially true for those areas adjacent to the brain's ventricular system. There is no doubt that destruction or damage to the hypothalamus and corresponding neuro-endocrine organs leads to potential developmental complications (physical and mental).*"

Dr. Louis Elsas, then director of Emory University School of Medicine, Department of Pediatrics, Division of Medical Genetics, testified before Congress on November 3, 1987, that he had spent 25 years *"trying to prevent mental retardation and birth defects caused by excess phenylalanine. And herein lies my basic concern, that aspartame is in fact a well known neurotoxin and teratogen which, in some as yet undefined dose, will both reversibly in the adult and irreversibly in the developing child or fetal brain, produce adverse effects."*

"Many studies of both acute and chronic ingestion of 34 mg aspartame/kg/day have demonstrated a two- to five-fold increase in semi-fasting blood phenylalanine concentrations (from approximately 50 to 250 μM) without concomitant increases in tyrosine or other amino acids. The degree of increase by normal humans depends on several variables including the efficiency of gut transport, liver utilization and growth rates. It was thought by many scientists that this degree of blood phenylalanine increase would not affect brain function. However, currently available information indicates that this is not true.

"In the developing fetus such a rise in maternal blood phenylalanine could be magnified four to six fold by the concentrative efforts of the placenta and fetal blood brain barrier. Thus, a maternal phenylalanine of 150 μM could reach 900 μM in the developing fetal brain cell and this concentration kills such cells in tissue culture. The effect of such an increased fetal brain concentration in vivo would probably be much more subtle and expressed as mental retardation, microcephaly, or potentially certain birth defects.

"In the rapidly growing post-natal brain (children of 0-12 months) irreversible brain damage could occur by the same mechanism.

"In the adult, we have found that changes in blood phenylalanine in these concentration ranges are associated with slowing of the electroencephalogram, and prolongation of cognitive function tests. Fortunately, these effects on the mature brain are reversible but provide clear evidence for a negative effect on sensitive parameters of brain function."

Difficulty walking

"The top of my feet have become numb and my walking gait has changed.... The only aspartame that I consume is from sugarless gum. I had been chewing up to 24 pieces per day. Is that enough to cause problems?"

"I am a 68 year old male. I have been a runner and then a walker for the last 26 years, until I developed a problem with my feet. The top of my feet have become numb and my walking gait has changed. I went to the doc, he did a brain and back scan. The results came back OK. I would walk 10 to 12 miles per day at a 14 minute pace, 6 days a week.... Since my walking gait has changed, my pace has slowed and I am not as steady as I had been.... On the web, I found a connection with walking gait change, foot drop and MS [multiple sclerosis]. The only aspartame that I consume is from sugarless gum. I had been chewing up to 24 pieces per day. Is that enough to cause problems?"

> *Note: According to the website aspartamekills.com, "In the May 1992 edition of their journal, Flying Safety, the United States Air Force warned all pilots to stay off aspartame, stating: 'Some people have suffered aspartame related disorders with doses as small as that carried in a single stick of chewing gum.'"*
>
> *According to Jim Bowen, MD, "Aspartame in chewing gum is absorbed directly though the buccal mucosa of the tongue, mouth, and gums, making it a far worse poisoning than even if it were given intravenously. The nerves serving this area and their vascular supply derive directly from the brain, so the aspartame absorbed through them goes directly into the brain, by-passing the spinal cord and blood brain barrier."*

The male on aspartame shown in Figure 4-3 had trouble walking and frequently fell over. His body leaned toward the left. He also appeared to have mild symptoms of torticollis. See "Spasmodic torticollis" on page 28.

While walking, this male on aspartame continually leaned toward his left.

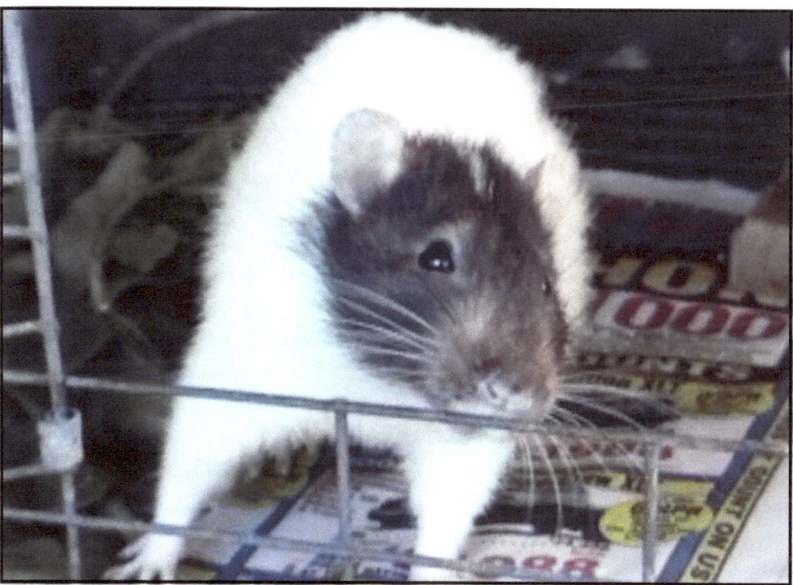

FIGURE 4-3: Male on aspartame with difficulty walking and mild symptoms of torticollis

The female on aspartame shown in Figure 4-4 also had trouble walking.

This female on aspartame also leaned toward her left as she walked.

Her leaning is so extreme in this photo, it looks as if she is falling.

FIGURE 4-4: Female on aspartame with difficulty walking

Eye disorders within my aspartame group

"For many years I had consumed four to five diet sodas daily. I am a jogger and very active with a pretty good diet.... In 2002 I had an eye infection and had to see an ophthalmologist who told me I had a fairly advanced cataract in one eye and the beginnings in the other. My father had them so I assumed that this was genetics. In August, 2004 I went back for a check-up and the ophthalmologist said that he felt that I should consider surgery on the left eye. I put it off. Then I received some of the material about aspartame from a friend.... As I read this stuff I could not believe how many of the symptoms they described fit me. After quitting aspartame for approximately two months I had my most recent eye exam in Aug. 05. The quote: 'Your eyes look fantastic.... no surgery needed.'"

The following subsections shows rats on aspartame with eye infections, bleeding, and protruding eyes.

Eye infections

According to Dr. Janet Starr Hull, when consuming aspartame you are more likely to have an increased susceptibility to infection. Dr. Roberts reports possible correlations between aspartame consumption and infection, including that associated with acquired immune deficiency syndrome (AIDS), chronic Epstein-Barr infection, rheumatic fever, and yeast infection.

Regarding the possible mechanisms involved, Dr. Roberts states that:

"Increased phenylalanine [a breakdown component of aspartame] *appears to alter cell-mediated immunity. This is evidenced by the enhanced immunity*

noted in both animals and humans placed on phenylalanine restriction. In turn, elevation of the serum phenylalanine by infection can contribute to a vicious cycle. The phenylalanine/large neutral amino acid ratio increases in acute infection by as much as 50 percent."

I observed two males and a female on aspartame with eye infections, as shown in Figure 4-5 through Figure 4-7.

This male and the female below developed eye infections.

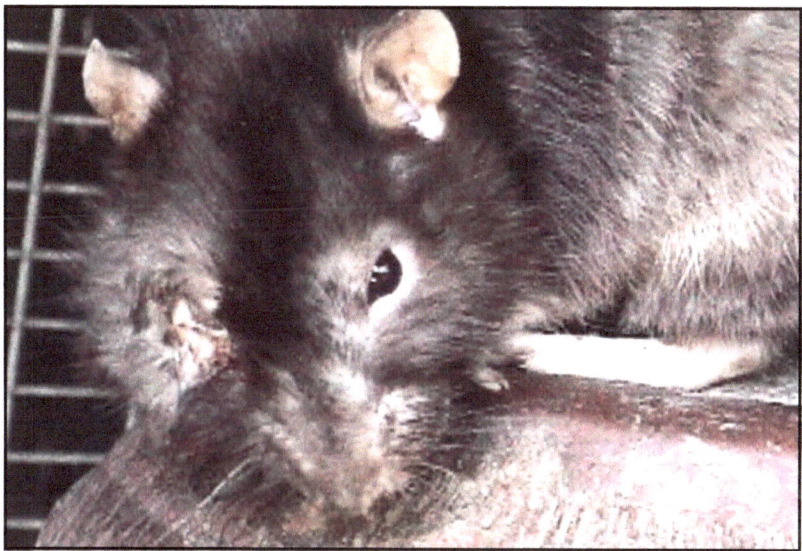

FIGURE 4-5: Male on aspartame with an infection in his right eye

The eyes of the female on aspartame in Figure 4-6 also became infected.

Note the eye puffiness and pool of blood under her left eye.

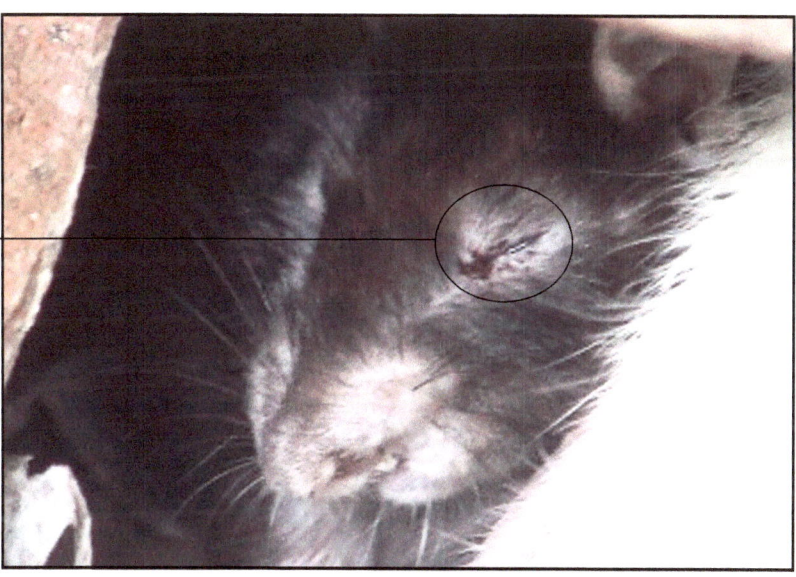

FIGURE 4-6: Female on aspartame with infected eyes

The female in Figure 4-6 also had bleeding eyes, a possible symptom of aspartame poisoning. See "Bleeding eyes" on page 35.

The male in Figure 4-7 developed an eye infection after a fight.

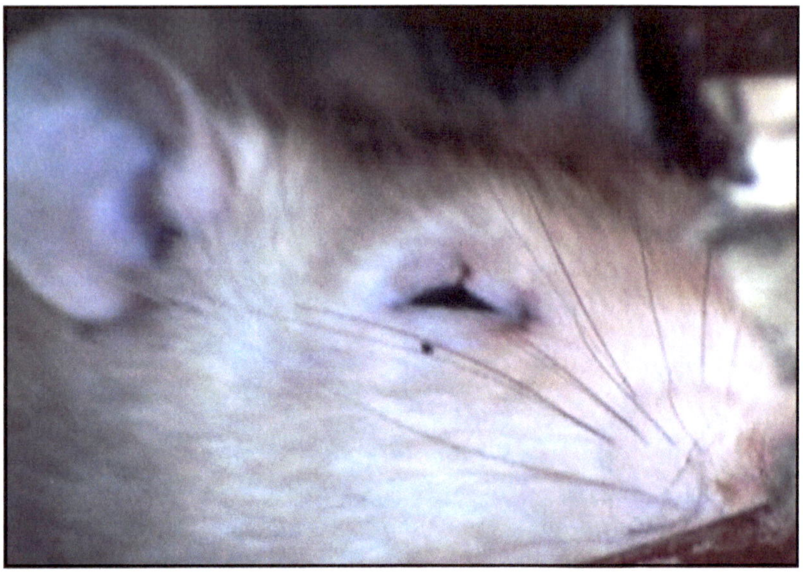

FIGURE 4-7: This male on aspartame also had an eye infection

> The rats on aspartame seemed significantly more aggressive than the controls.

Speaking of fighting, the rats on aspartame seemed significantly more aggressive than the controls. I once saw two males on aspartame on their hind legs fighting violently. The controls were relatively passive. According to the *National PKU News*, a high blood level of phenylalanine, the majority breakdown (50%) component of aspartame, may induce aggressive behavior in people with phenylketonuria (PKU), an inborn disorder of phenylalanine metabolism often associated with mental retardation in children. The tendency toward aggressive behavior is elucidated by a concerned teen in the Q&A section of the *National PKU News:*

"Q. Is there a relationship between PKU and aggressive tendencies? I am a teen who has a problem with becoming extremely angry and aggressive, almost violent at times. Is this common, or is it just me?

"A. This behavior suggests to me that you are either off the [PKU] *diet, or if you are on the diet that your blood phe* [phenylalanine] *levels are too high. PKU patients who are off diet or whose levels are too high are often aggressive and less stable."*

Several studies cited in an article entitled "Behavioral and Neurological Effects of Aspartame," claim that a diet excessive in phenylalanine *can induce PKU*, and can therefore induce aggressive behavior. Such behavior may be explained by how phenylalanine is processed by the blood-brain barrier before entering the brain. When ingested, digested, and carried to the blood-brain barrier, phenylalanine competes with the amino acid tyrosine to become attached to the neutral amino acid that transports it to the brain. When excess phenylalanine is consumed, the brain gets less tyrosine, found in high concentrations in meats, whole grains, dairy, avocados, bananas, legumes, beans and nuts. After entering the brain, tyrosine is converted into norepinephrine and dopamine. Norepinephrine is a mood elevator that stimulates a sense of well being, and its depletion can lead to depression, which can in turn, lead to aggression. In a study of 42 normal women and 23 normal men, aggression and depression were found to have a significant positive correlation for the women, but not the men. A study of 1101 persons with dementia found those manifesting physical or verbal aggression had a higher prevalence of depression. When 2083 Norwegian pupils in Grade 8 were surveyed, a significant positive correlation was found between depression and aggressive, bullying behavior. An analysis of surveys taken for 41 eleven-year-olds and 22 fifteen-year-olds found significant correlations between depression and aggression in the 11- year-olds, and an even stronger correlation among the 15-year-olds.

> *Note: The depletion of dopamine due to the over-consumption of phenylalanine can lead to Parkinson's disease. Several drugs are sold to treat the disease by increasing dopamine levels in the brain. Dr. Blaylock began a 20-year quest for the causes of Parkinson's after his father died from it in March, 1989. Blaylock found that Parkinson's, Alzheimer's, and Lou Gehrig's diseases may be correlated with the use of aspartame, MSG, and other neurotoxic chemicals he discusses in Excitotoxins: the Taste that Kills.*

The saturation of phenylalanine from aspartame consumption can also lead to the depletion of the amino acid tryptophan, a precursor of serotonin, thought to be involved in the control of appetite, sleep, memory, learning, temperature, mood, behavior (including sexual and hallucinogenic), the cardiovascular system, muscle contractions, the endocrine system, and depression.

When aspartame is consumed over time, the resulting depletion of tryptophan can cause depression and aggressive behavior due to a lack of serotonin in the brain and spinal fluid.

Phenylalanine, tyrosine, and tryptophan are classified by biochemists as aromatic amino acids because of their chemical structures. As such, they compete with one another to pass through the blood-brain barrier. When aspartame is consumed over time, the resulting depletion of tryptophan can cause depression and aggressive behavior due to a lack of serotonin in the brain and spinal fluid. Three studies done by the University of Texas using 24 men, 8 men, and 12 women, respectively, found that when subjects were provoked, aggressive incidents increased under tryptophan-depleted conditions and decreased when blood levels of tryptophan were elevated.

Six other studies have correlated serotonin depletion with impulsive and violent criminal behaviors. Six more have associated the condition with alcohol abuse and dependence. Three studies have correlated a lack of serotonin with Gilles de la Tourette's syndrome. Two studies associated a lack of serotonin with bulimia. Four studies have associated serotonin depletion with suicide attempts, and three have associated it with children institutionalized for aggressive behavior.

Bleeding eyes

In addition to bleeding in the eyes of the female in Figure 4-6, the eyes of the male on aspartame shown in Figure 4-8 were visibly bleeding.

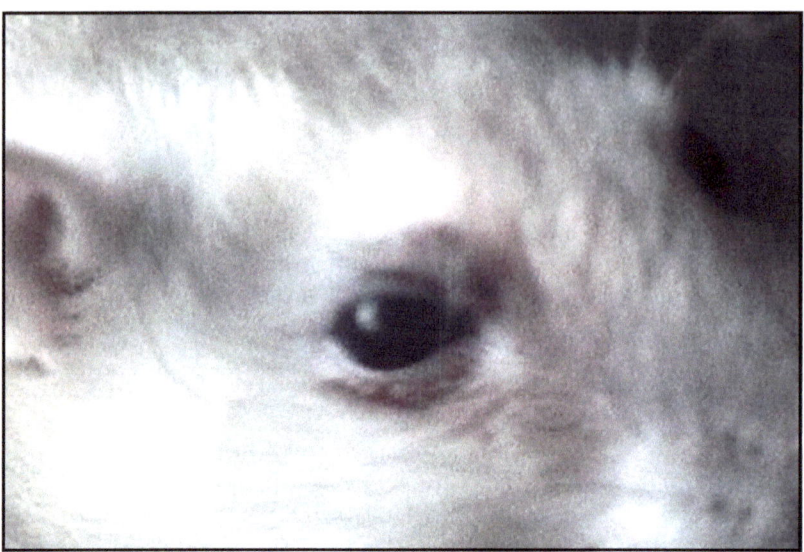

FIGURE 4-8: This male on aspartame had bleeding eyes

The eyes in the female on aspartame shown in Figure 4-9 were infected and bleeding. This female also had a mammary or lymph gland tumor as shown in Figure 3-8.

The eyes in the female on aspartame shown in Figure 4-9 were infected and bleeding.

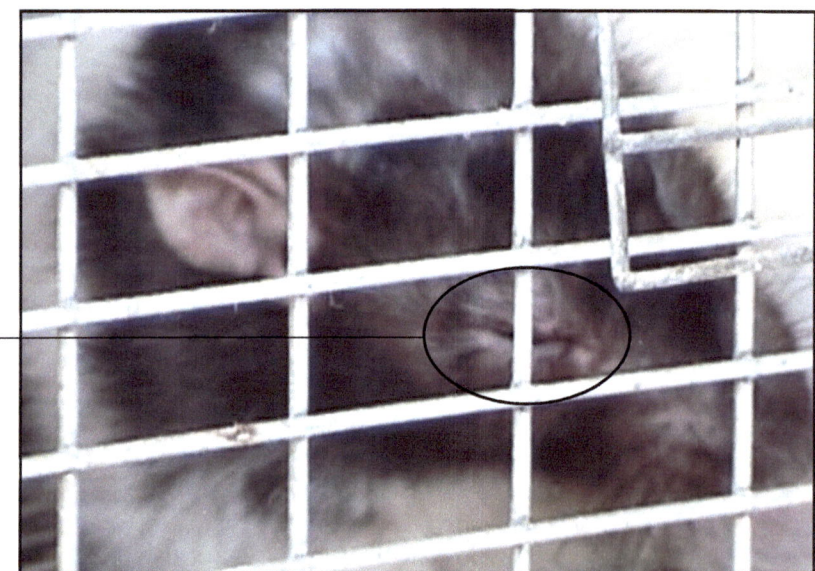

Bleeding eye ———

FIGURE 4-9: This female on aspartame had bleeding eyes

Bleeding eyes have also been found to occur in humans consuming aspartame. According to psychiatrist Ralph Walton, MD, *"In our double-blind study here at this hospital, we had a really tragic situation which occurred, which I attributed directly to the aspartame. We needed volunteers. We looked at both patients, that is people who had a history of a mood disorder. And we needed some controls, that is, people without a history of mood disorder.*

"Several days into the study he had sudden bleeding in his eye and a detachment of his retina."

One of the people that I used [in the aspartame group of] *the study was the administrator for our psychiatric staff with a PhD in psychology. Several days into the study he had sudden bleeding in his eye and a detachment of his retina, and had to be rushed to Cleveland for emergency surgery. His eye could not be saved. He lost the vision in one eye. At the same time, we had another participant in [the aspartame group of] this study—a nurse—who also had bleeding in her eye. So we had two people who during the course of the study had eye emergencies."* Dr. Walton subsequently discontinued the study.

Protruding eyes

The female on aspartame shown in Figure 4-10 had protruding eyes—a symptom of the thyroid disorder called Grave's disease in humans.

According to Dr. Janet Starr Hull, *"In 1991, I was diagnosed with an 'incurable' case of Grave's Disease, a fatal thyroid disorder. I never really had Grave's Disease but my doctors were convinced I did. I had aspartame poisoning with symptoms of 'textbook' Grave's Disease caused by aspartame saturating my foods."*

FIGURE 4-10: This female on aspartame had protruding eyes

Dr. Hull continues:

I had protruding eyes, cystic acne, and my hair was falling out in clumps. I had gained 30 pounds, too. ALL that went away within a year from stopping all aspartame. I even had holes in my retina from the methanol, and those have closed up now. All backed up by my eye surgeon."

The females on aspartame shown in Figure 4-11 and Figure 4-12 also had protruding eyes.

Protruding eyes are a symptom of the thyroid disorder called Grave's Disease

FIGURE 4-11: **Females on aspartame with protruding eyes**

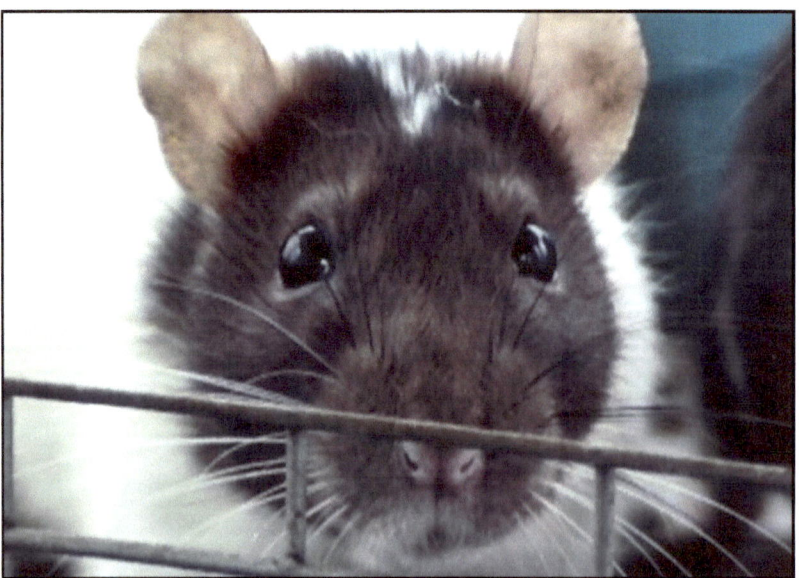

FIGURE 4-12: **Female on aspartame with protruding eyes**

Blindness

"Several years ago I used to consume about one roll of aspartame-containing mints per day. After having an eye exam where the doctor performed the 'air jet' test, I developed obliterative vasculitis. The tiny veins in my eyes were breaking and bleeding, which nearly blinded my left eye."

"Several years ago I used to consume about one roll of aspartame-containing mints per day. After having an eye exam where the doctor performed the 'air jet' test, I developed obliterative vasculitis. The tiny veins in my eyes were breaking and bleeding, which nearly blinded my left eye. (I quit those mints MANY years ago but should have done so sooner). I then went to another eye doctor who performed laser eye surgery and completely healed my right eye. The left eye was mostly healed, but now I suffer from macular degeneration or warping in that eye. It is like having a movable black zone or blind spot, but fortunately my right eye has been able to compensate. The vision tester at the DMV thought I was completely blind in that eye."

I was unable to get video, and it probably would have been difficult to see in a photo, but one of my females on aspartame acted as if she were blind. Blindness has also been reported as one of the adverse effects of methanol, a breakdown component of aspartame.

Skin disorders within my aspartame group

By 1987 the FDA had already received over 3,600 consumer complaints about aspartame. The Aspartame Consumer Safety Network reported in 1996 that they had filed over 10,000 consumer complaints with the FDA. The adverse effects attributed to aspartame in complaints submitted to the FDA were published in April 1995 by the FDA's parent organization, the Department of Health and Human Services (HHS). The data was obtained by a Freedom of Information Act (FOIA) request by reporter Barbara Mullarkey. It included a list of 92 symptoms reportedly resulting from aspartame[51] that shows 114 people reporting skin problems.

Figure 4-13 shows an open ulcer that never healed, found above the right hind leg of a male on aspartame.

Open ulcer on a male on aspartame.

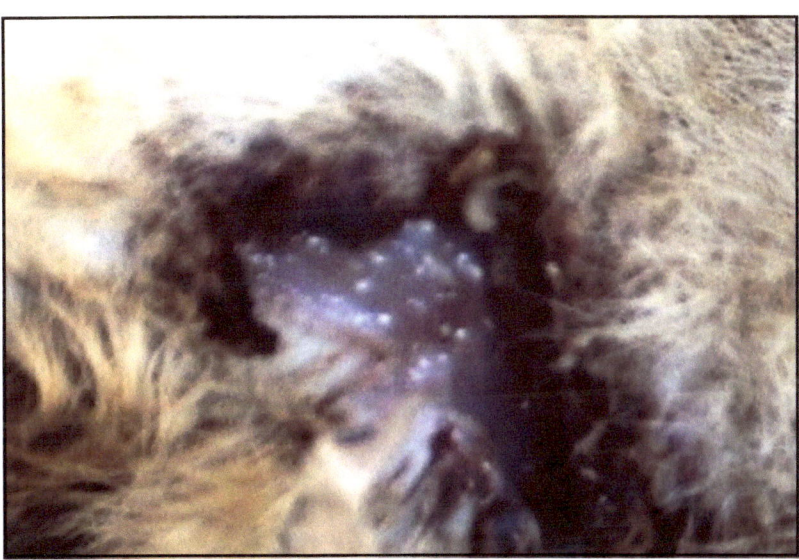

FIGURE 4-13: Open ulcer that never healed on a male on aspartame

These males on aspartame developed skin lesions.

The males on aspartame in Figure 4-14 through Figure 4-17 had skin lesions.

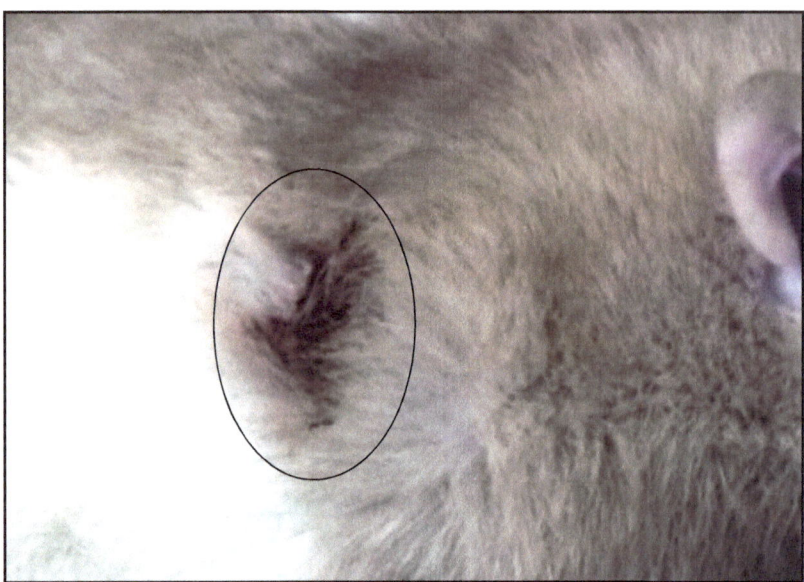

FIGURE 4-14: Male on aspartame with a skin lesion

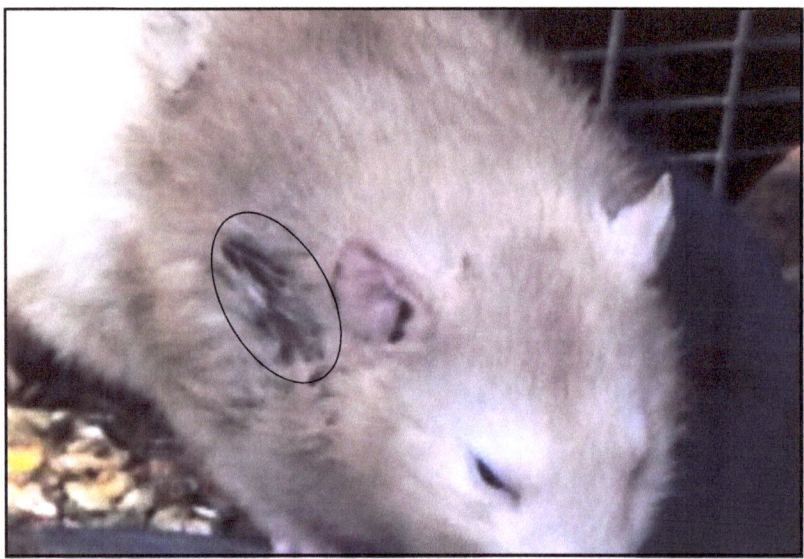

FIGURE 4-15: Another male on aspartame with a skin lesion

These males
on aspartame
also developed
skin lesions.

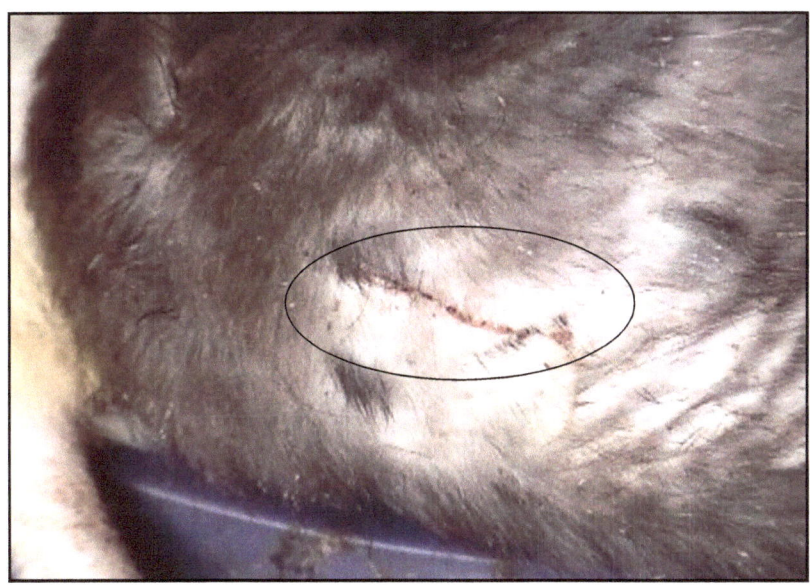

FIGURE 4-16: **Another male on aspartame with a skin lesion**

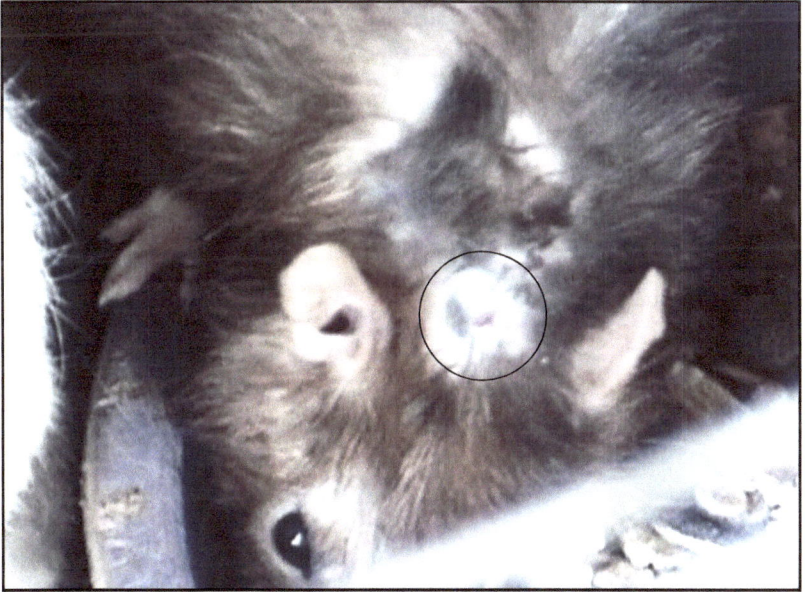

FIGURE 4-17: **Male on aspartame with a skin lesion**

The male in Figure 4-17 had a skin lesion and thinning fur. Also see "Thinning fur" on page 42.

About one-third of the skin on the back of the aspartame female shown in Figure 4-18 became separated from her body about a week before she died.

The skin of this female started coming off a week before she died.

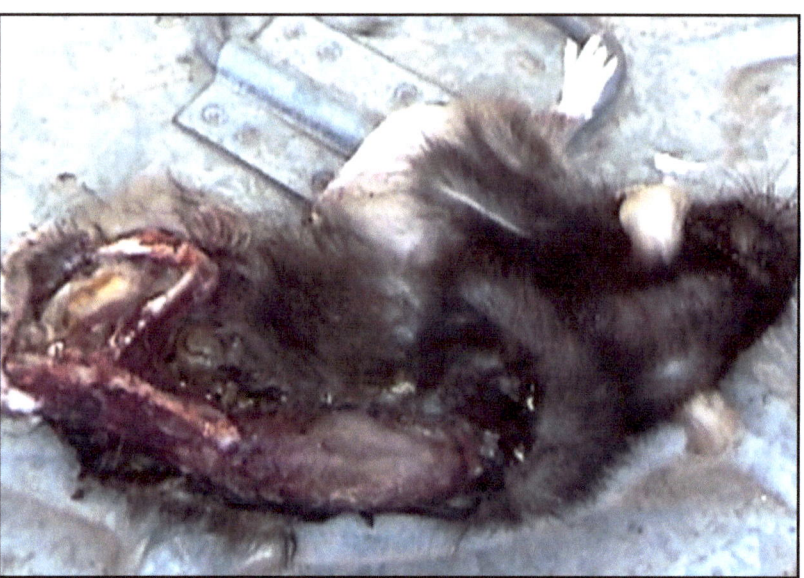

FIGURE 4-18: Female on aspartame with a severe skin problem

Thinning & yellowing fur within my aspartame group

The following subsections describe rats with thinning and yellowing fur.

Thinning fur

"I was buying 12 packs of Diet Coke.... I would grab a 12 pack every other day.... On March 11th, 1999 ... I was feeling very sick. I hadn't felt good for a few years now. I was always tired. Full of aches and pains. Felt like I was ready to have my 90th birthday. That is when I realized how much Diet Coke had taken over my life. So what symptoms did I have? ... Thirst - Weight gain (50 lb.) - Tired all the time - Aches in my joints - Throbbing headaches - Hair loss - Blurred vision - Mood swings - Depression - Couldn't think straight - Lived in a haze - Cramps - Rashes - Numbness in my legs & arms - Confusion....

"My mother ... was worried about my consumption of Diet Coke.... Mother had heard about ... how aspartame was bad for you and caused brain cancer. Well ... I shrugged it off.

"By the time I figured out what was causing it, I had lost about half my hair volume."

"The hair loss caused me great pain. As a teenager I had the thickest most beautiful hair. I would comb my hair and have a sink full of it. Just running my hands through my hair would produce a handful. It fell out all over the place.... By the time I figured out what was causing it, I had lost about half my hair volume.

"On March 11th, 1999 ... I was talking to my boyfriend and for some reason I picked up my Diet Coke can. The word ASPARTAME stuck out at me.... So I

typed in aspartame into Excite and I found www.aspartamekills.com.... I dumped my Diet Coke out and I have not touched it since.

"All those symptoms above I listed. They are all gone. Every one of them. My hair is even growing back. I can comb my hair without a ton of hair on my brush instead of my head. My hair stopped falling out within a week. It is amazing how I feel."

The males on aspartame shown in Figure 4-19 and Figure 4-20 had thinning fur.

"All those symptoms above I listed. They are all gone. Every one of them. My hair is even growing back. I can comb my hair without a ton of hair on my brush instead of my head. My hair stopped falling out within a week. It is amazing how I feel."

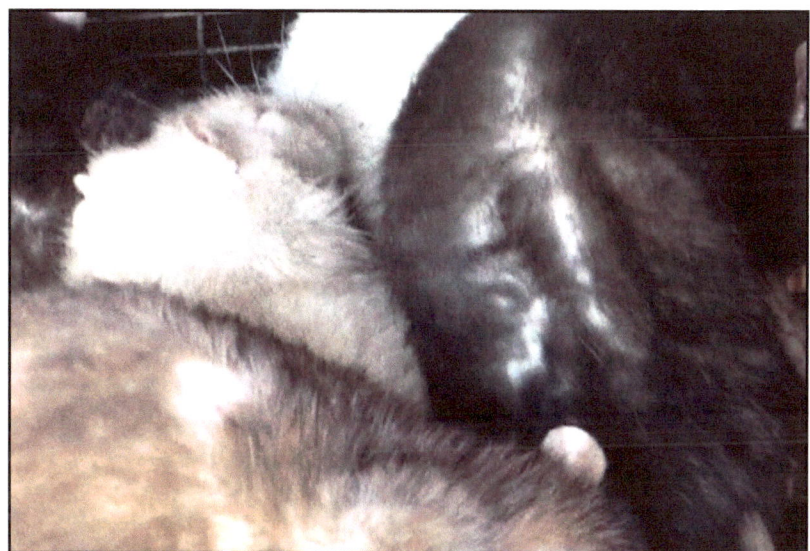

FIGURE 4-19: **Aspartame males started losing their fur.**

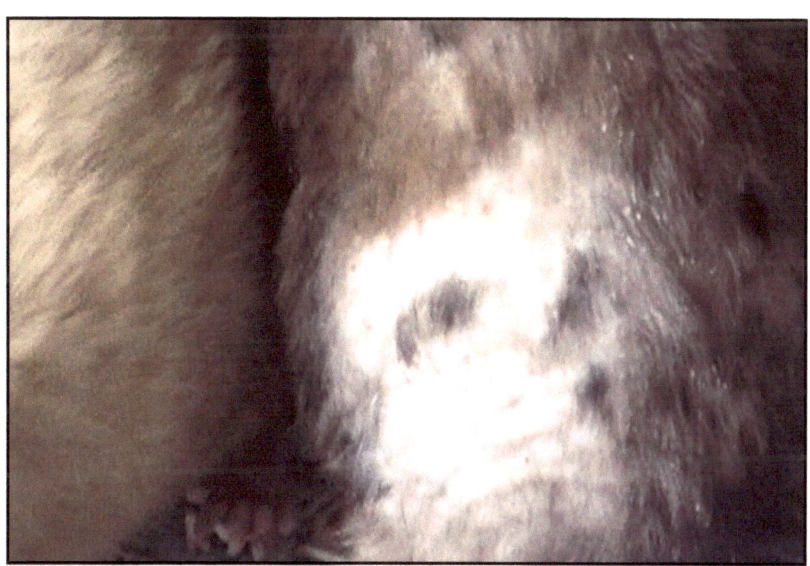

FIGURE 4-20: **Aspartame male losing his fur.**

A few of the rats in the control group also developed thinning fur, so I thought it may be part of the normal aging process. However, some suspect that aspartame is the cause of their hair loss. Recall that Dr. Janet Hull complained of "*hair falling out in clumps*" in the quote on page 37, and here's another person's experience:

"*I am a 40 year old woman.... Until six months ago I had glorious, thick auburn hair. Now, I have the horseshoe patterns with tufts of hair on top, as is frequently seen in balding men.*

"*I have always consumed 3 to 6 diet sodas per day, and have never noticed any problems with my hair until recently.... About a year ago I decided to make a conscious effort to eliminate all refined sugar and high fructose corn syrup from my diet. This eliminated not just any candy, cookies, ice cream, etc., but also most fruit juices (read the labels!) and also, fruit flavored yogurt and fruit preserves/jams/jellies. I replaced these items in my diet with 'Lite' versions, which contain NutraSweet (aspartame)....*

"*For a few months, I noticed that my scalp felt 'tight.' ... Then, the hair started falling out in clumps and within the passing of two months, I had lost more than half of the hair on my head! My scalp is now clearly visible over the top and sides of my head, and the hair in the back of my head is noticeably (to me) thinner. Any time I touch my head, my hand comes away with 10 or more hairs, which are very thin at the scalp end (the recent growth of 1/4 inch is less than half the thickness of the rest of the strand) and the root is still attached when the hair falls out. Does anyone else suspect their alopecia is due to use of NutraSweet?!?!?!?*

> "[My] *hair started falling out in clumps and within the passing of two months, I had lost more than half of the hair on my head!*"

Yellowing fur

According to the book *The Rat*, yellowing fur can be an effect of the natural aging process. The World Health Organization, however, reports that yellowing fur can be a symptom of formaldehyde exposure. Formaldehyde is a breakdown component of the methyl-alcohol component of aspartame. Formaldehyde has been shown to collect and persist in vital organs after the consumption of aspartame. According to the Trocho study:

"*The chronic treatment of a series of rats with 200 mg/kg of non-labelled* [non-radioactive] *aspartame during 10 days resulted in the accumulation of even more label* [radioactive formaldehyde] *when given the radioactive bolus* [a very large amount of radioactive aspartame administered at a single time], *suggesting that the amount of formaldehyde adducts* [the accumulation of formaldehyde bound to proteins] *coming from aspartame in tissue proteins and nucleic acids may be cumulative. It is concluded that aspartame consumption may constitute a hazard because of its contribution to the formation of formaldehyde adducts.*"

> ***Note: My understanding of this quote is that when the body is still metabolizing non-radioactive formaldehyde that has accumulated from previous doses, and is then given a large amount of radioactive aspartame at a single time (a bolus), then more of the radioactive formaldehyde accumulates in the body because the non-radioactive formaldehyde is still being metabolized.***

Formaldehyde adducts are difficult to eliminate from the body. They damage the nervous and immune systems and cause irreversible genetic damage.

Formaldehyde is often found in dark wines. A few years ago, a friend became the caregiver for her 80-something year-old father who had been given copious amounts of wine by his previous caregiver, and his hair had yellowed. When he stopped drinking the wine, the yellowing disappeared.

In my experiment, the females on aspartame shown in Figure 3-4 through Figure 3-13 and the males on aspartame shown in Figure 3-21 and Figure 3-22, appear to have yellow tinges or patches on their otherwise white fur. The white and black control females shown in Figure 3-28 through Figure 3-31 also appeared to have yellowing fur, though in general the yellowing appeared to be significantly less than on the fur of the white and black females in the aspartame group.

> *Note: I was unaware of the issues surrounding yellowing fur until writing this report, and therefore did not look for it on my rats during the experiment. So there may have been others with yellowing fur that I did not photograph. I apologize that the yellowing cannot be seen in the black-and-white version of this report.*

Obesity within my aspartame group

"Consuming aspartame seems to trigger a hunger type feeling, even if I have just eaten, whereas if I abstain from aspartame, I would be satisfied. Does this have any scientific validity, or am I just imagining it? I really get ravenous after consuming 'straight' aspartame like in a diet coke, versus aspartame added to a food, like a milkshake or oatmeal sweetened w/ aspartame."

During my experiment, the female on aspartame in Figure 4-21 grew obese. The male with a tumor shown in Figure 3-25 also grew obese.

This female grew fat.

None of the rats from the control group became fat.

Ironically, aspartame is sold to help people lose weight.

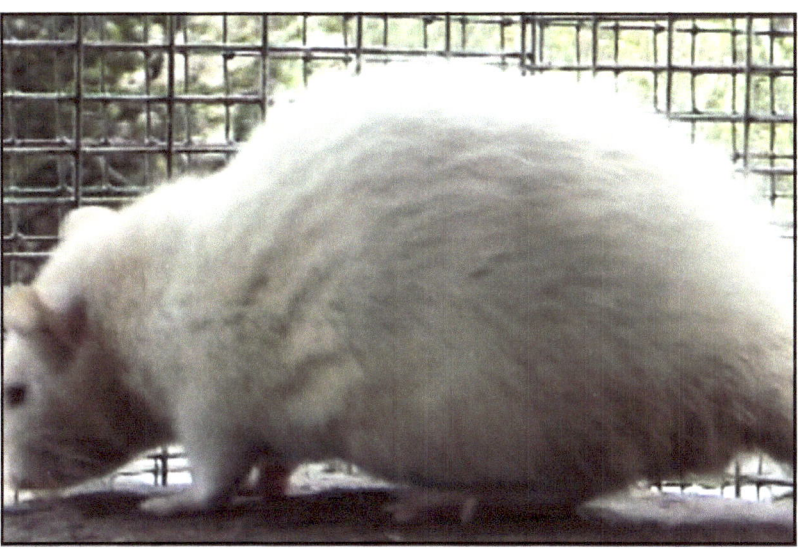

FIGURE 4-21: Female on aspartame that became obese

For more about aspartame, MSG, and obesity, and for my own personal story, see "Does Aspartame Make You Fat?" on page 53. "Does Aspartame Make You Fat?" in *My Aspartame Experiment: Report from a Private Citizen*, available from aspartameexperiment.com.

What could be worse than a diet product that makes you fat? One that is also addictive! See "Is Aspartame Addictive?" in *My Aspartame Experiment: Report from a Private Citizen*, available from aspartameexperiment.com.

Miscellaneous adverse effects in my control group

All rats in my control group were free of visible symptoms of neurological damage. Some males from the control group developed thinning fur. One female from the control group had skin problems.

All rats in my control group were free of visible symptoms of neurological damage. Some males from the group developed thinning fur. One female from the group had skin problems; however, the damage appears to have occurred after death (see Figure 3-31 on page 26). Some females from the control group appeared to have yellowing fur, but not as yellowed as the fur of the rats in the aspartame group. See "Yellowing fur" on page 44.

When a preliminary version of this report was summarized and leaked on the Internet without my permission, it caused a stir on blogs reddit.com and digg.com. See "Misinformation About My Experiment" in *My Aspartame Experiment: Report from a Private Citizen*, available from aspartameexperiment.com. One blogger commented that my control group was too healthy compared to those from other studies. The person claimed that I must have faked my data. See "Did I Fake My Data?" in *My Aspartame Experiment: Report from a Private Citizen*, available from aspartameexperiment.com.

I had actually been concerned when my controls started developing tumors, and investigated possible causes such as pesticides and herbicides in their food. See my frightening findings in "What may have caused my control group tumors" in *My Aspartame Experiment: Report from a Private Citizen*, available from aspartameexperiment.com. After being accused of falsifying my data, I researched possible reasons my control group may have been healthier than those in other studies. See "Why was my control group relatively healthy?" in *My Aspartame Experiment: Report from a Private Citizen*, available from aspartameexperiment.com.

5

Conclusion

"*[Aspartame] does not produce adverse effects, even at doses several orders of magnitude greater than human consumption rates.*"

"*Prior to the marketing of aspartame, numerous studies were done to evaluate its metabolism and tolerance in healthy subjects and various subpopulations. Postmarketing, numerous additional studies were done, including studies to evaluate alleged sensitivity to aspartame. ...* [These studies indicate that aspartame] *does not produce adverse effects, even at doses several orders of magnitude greater than human consumption rates.*"

—Dr. Harriett H. Butchko, MD, et. al., former employees of Monsanto Company, The NutraSweet Company, and G.D. Searle and Company; previous manufacturers of aspartame

This chapter presents a summary of the data I observed during my experiment. For more information, see *My Aspartame Experiment: Report from a Private Citizen*, available from aspartameexperiment.com.

Summary of my experimental results

Out of the 108 rats in my study, 30 males and 30 females—a total of 60 rats—received aspartame (now rebranded as AminoSweet[3]) in the form of NutraSweet mixed in their drinking water. The control group consisted of 24 females and 24 males—a total of 48 rats.

Tumors in my aspartame group

The photos in "Resulting Tumors" starting on page 5 show:

A total of 20 females from my aspartame group developed visible tumors.

That's 67% of all my females on aspartame.

- A total of 20 females from my aspartame group developed visible tumors. That's 67% of all my females on aspartame.

- Seven males from my aspartame group developed visible tumors. That's 23% of all my males on aspartame.

The total number of rats—both male and female—in my aspartame group with visible tumors was 45%.

Possible reasons for the large number of tumors in my females on aspartame are analyzed in "Why so many tumors in my females on aspartame" in the book *My Aspartame Experiment: Report from a Private Citizen*, available from aspartameexperiment.com.

Tumors in my control group

Five females from my control group developed visible tumors . That's 21% of my female controls. No males from my control group developed visible tumors, so the total number of tumors in my control group was 5, or 10%. Tumors in my aspartame group generally grew much larger than those in the control group. Until I learned that most if not all rat experiments result in tumors in their control groups, I was concerned about the tumors within my control group. For a frightening analysis of why my control group females may have developed tumors, see "What may have caused my control group tumors?" in the book *My Aspartame Experiment: Report from a Private Citizen*, available from aspartameexperiment.com.

Other adverse effects in my aspartame group

In addition to tumors, the rats in my aspartame group developed other health problems, as described in "Other Adverse Results" starting on page 27. These results include:

- Four rats with apparent neurological disorders

- Eight rats with eye disorders

- Six rats with skin disorders.

- Two cases of thinning fur, and 12 cases of yellowing fur

Other adverse effects in my control group

In addition to the tumors shown in the previous chapter, my control group developed other health problems, as described in "Miscellaneous adverse effects in my control group" on page 46. These results include:

- Three males with thinning fur.

- One female had skin problems; however, the damage appears to have occurred after death.

Did my experiment demonstrate adverse effects of aspartame?

Though I did not have necropsies performed on the rats in my study, I believe that a significant number of the tumors I observed were indeed cancerous.

Though I did not have necropsies performed on the rats in my study, I believe that a significant number of the tumors I observed were indeed cancerous. I also believe that there were cancers that I did not see. I base these opinions on necropsies I had done on rats from a multi-generational aspartame experiment I undertook after completing the experiment documented in this report. I stopped the second experiment midway because it was too difficult to do on my own, it got out of hand, and I realized I could not provide adequate data to justify continuing.

During that experiment, I had three rats on aspartame necropsied by the county veterinarian. (I was unaware of that service during my previous experiment.) Of the rats I had necropsied, one had visible tumors and the others appeared to be in good health before their deaths.

The tumors I observed on one of the rats—a black and white male—did in fact turn out to be cancerous. Malignant tumors were also found in both its armpits and liver that I had not observed.

Here are the diagnoses from the pathologist:

- *"Malignant lymphoma, submandibular* [cancer of the lymph gland under each side of the lower jaw; each with a diameter of 5 cm = 2 in, which I observed] *and axillary lymph nodes* [in each armpit; each with a diameter of 2 cm = 0.8 in, which I did *not* observe.]

- *Fibrosarcoma, skin on top of head* [a malignant tumor of fibrous connective tissue, with a diameter of 2.5 cm = 1 in, which I observed]

- *Hepatoma, focal, liver* [tumor of the liver, with a diameter of 3 mm = 0.12 in, which I did *not* observe]*"*[4]

The two rats that appeared to be healthy were females. One was found to have malignant tumors. Her pathology report shows these diagnoses:

- *Chondrosarcomas (lung)* [cancerous tumors, each with a diameter of 0.1 - 0.3 cm = 0.03 - 0.1 in]

- *Fibroadenoma (mammary gland)* [cancerous tumor measuring 4 x 2.5 x 2 cm^3 = 1.6 x 1 x 0.8 in^3]

The pathologist made the following comment about this female:

"The large tumor within the lung has formed multiple metastases within the lung itself. The inflammation in the surrounding lung tissue is due to compression and obstruction by the tumors. Evidence of infectious pulmonary diseases (bacterial or viral) is not observed."

No cancer was found in the second rat that appeared healthy. Cause of death was reportedly meningitis. Her kidneys, however, were found to have "*acute tubular epithelial necrosis,*" a possible symptom of aspartame poisoning.

As described in "Chemical description" on page 2, about 10% of the aspartame molecule breaks down into methanol. Methanol is quickly oxidized into formaldehyde—a known carcinogen, and then formic acid, a cause of kidney necrosis and acute kidney failure. Formic acid appears to be the principal cause of adverse effects associated with methanol poisoning.

Is the FDA ADI for aspartame safe?

The FDA claims it is safe for a 150 lb. (68 kg) human adult (male or female) to drink up to twenty 12-oz. cans of diet soda per day. The acceptable daily intake (ADI) set for the European Union (E.U.) is 40 mg/kg, or 80% of the U.S. ADI, so a 68 kg human adult in the E.U. could consume up to 16 cans of diet soda per day.

The equivalent that a 150 lb. human male would receive in my experiment is about 13 cans of diet soda per day, which is 65% of the U.S ADI and 80% of the E.U. ADI. The equivalent that a 120 lb. human female would receive in my experiment is about 14 cans per day, or 70% of the U.S. ADI and about 90% of the E.U. ADI.

According to the pro-aspartame authors of *The Clinical Evaluation of a Food Additive: Assessment of Aspartame*:

"The Food and Drug Act of 1906 requires that food should not contain 'any added poisonous or other added deleterious ingredient which may render such article injurious to health.' ... Accepting that no substance can be shown to be absolutely safe, the objective of safety assessment is to determine the dose of a substance at which there is reasonable certainty of no harm. To accomplish this, FDA requires extensive animal toxicity studies. From these studies, the no-observed-effect level (NOEL) or no-observed-adverse-effect level (NOAEL) is determined. To ensure reasonable certainty of no harm, FDA normally sets the ... ADI at 1% of the NOEL or NOAEL... If the additive is found to cause cancer in either animals or humans at any dose, it is banned from use as a food additive as a result of the Delaney Anticancer Clause of 1958."

By FDA mandate, therefore, the U.S. ADI is supposed to be one-hundred times lower than the smallest amount that causes any adverse effect. Due to the evidence documented in this report, it appears to me that both the U.S. and E.U. ADIs for aspartame are too high. In fact, it appears to me that there is no safe level of aspartame consumption. For example, if you consider the aggregate of the grossly observed adverse effects documented in this report, I believe it would be fair to state that the levels of aspartame my rats received were *"injurious to health."*

The amount of aspartame approved for use by the FDA under its own guidelines should therefore be at most 1/100 of the amount received by my rats. My male rats received about 34 mg/kg/day and my females about 45 mg/kg/day. The ADI for human males should therefore be less than 0.34 mg/kg/day. For human females, the ADI should be less than 0.45 mg/kg/day. The maximum amount of soda consumed by human males and females under these proposed aspartame ADI levels should therefore be at most 1/7 of a can of soda per day.

Is further investigation of aspartame warranted?

Regarding the results of my experiment, I suggest that if they are ignored and the aspartame ADI remains the same, an experiment should be undertaken where diet drinks are administered at the equivalent levels given my rats to the following subjects for the rest of their lives: The officials who got aspartame approved for public use in 1981, including Donald Rumsfeld, ex-Secretary of Defense, and then President and CEO of G.D. Searle, who used political pull to get the chemical approved in the early months of the Reagan administration; Dr. Arthur Hull Hayes, who single-handedly approved aspartame at the FDA in 1981 over the objections of FDA scientists; those at the FDA who over the years have refused to ban aspartame from the marketplace, though it received 77% of all complaints registered by that organization; the boards of directors and officers of all aspartame manufacturers worldwide; the boards of directors and officers of all pro-aspartame organizations worldwide; and the boards of directors and officers of all manufacturers adding aspartame to more than 6,000 consumables worldwide.

www.ingramcontent.com/pod-product-compliance
Lightning Source LLC
Chambersburg PA
CBHW041508280526
45792CB00004B/1180